On

the general, the special and the general-special relativity theory

Dr. rer. pol. Erik Kolek

On the general, the special and the general-special relativity theory

Chronicles of Business Informatics Physics (CBIP)

Volume 1, edition no. 1.1

2024

Imprint

Scientific citation:

Kolek, Erik (2024). On the general, the special and the general-special relativity theory. In: *Chronicles of Business Informatics Physics (CBIP)*. Volume 1, edition no. 1.1. ISBN: 9783759712189.

Edition no. 1.1 contains didactic text improvements and more illustrations.

The Chronicles of Business Informatics Physics (CBIP) consist of cited significant scientific papers and the individually published research articles of the author Dr. rer. pol. Erik Kolek. As an author and an editor of the Chronicles of Business Informatics Physics (CBIP), Dr. rer. pol. Erik Kolek (erster.kontakt@erikkolek.de) is also responsible for editing, translation, typesetting, design (including cover), texts, images and cover picture. The book is produced and published by BoD – Books on Demand, Norderstedt, Germany.

Bibliographic information of the German National Library: The German National Library lists this publication in the German National Bibliography; detailed bibliographic data is available on the Internet at dnb.dnb.de.

Foreword

This book describes the content of Albert Einstein's special and general theory of relativity. This is followed by a derivation of a supplementary, completed theory of relativity, which is referred to in this work as the general theory of relativity. Overall, this book is a generally understandable treatise in the field of business informatics on the one hand and physics on the other. It therefore takes a holistic view of the theory of relativity. The structure of the content has been reorganized according to the derivation, starting with the general and special theory of relativity by Albert Einstein and ending with the general-special theory of relativity by Erik Kolek. In all theories, the law of the constant speed of light c is always at the forefront of physical considerations. This book is aimed at the general public.

The original work "On the special and general relativity theory" was written by Albert Einstein in 1916 and published just one year later. Since then, Albert Einstein has only changed the contents of the original work in the appendix after the photographs of the solar eclipse of May 29, 1919, because these snapshots of a moving body confirmed his general theory of relativity. Since this book is an extended (generally) understandable treatise in the field of business informatics physics, this book therefore cites a reprint from 2009 and only this one reference by Albert Einstein.

This book provides a simple yet accurate introduction to Albert Einstein's general and special theories of relativity and Erik Kolek's general-special theory of relativity. To understand this book, a basic knowledge of the mathematics of theoretical physics is required, as the contents are described based on a general scientific and philosophical view of this theory of relativity.

The knowledge of Albert Einstein is also conveyed in an easily understandable way by means of supersymmetric and relativistic modelling and visualization. The model-based representation of the universe as well as sections of it enables readers, regardless of whether they have studied physics, to better understand Albert Einstein's multidimensional way of thinking. Due to this necessary subjective deductive way of thinking, the content structure of Albert Einstein's work must be observed so that no changes in content can occur and because the aim of this book is to convey the general and special theory of relativity once again in a way that is generally understandable and, in addition, as simplified as possible. Building on this, a further, more complete, i.e. an extended theory of relativity is introduced, which is easy to find by means of supersymmetric and relativistic modelling and visualization and which enables new general-specific views of the entire universe.

Alternative matrix. The contents structured in this book resemble a derivation from the general, via the special, to the general-special theory of relativity. If readers would like to experience a different thought matrix according to the origin of the connections, I recommend a classical organization of this treatise according to Albert Einstein for the order of reading: Third section: On the special relativity theory, Second section: On the general relativity theory, First section: On the foundations of the general and special relativity theory, Fourth section: On the general-special relativity theory, and finally Fifth section: Appendix.

Abstract reading note: The sometimes long content sentences each describe a figurative idea necessary for understanding, so readers are asked to read these conclusions with a little patience and diligence.

May 2024. Dr. rer. pol. Erik Kolek, Diplom-Betriebswirt (FH), M.A., M.Sc.

Table of Contents

First section: On the foundations of the general and special relativity theory

1 Comprehensible derivation (deduction) of the Lorentz transformation

If the relative alignment of two coordinate systems (reference systems) is constantly in relation to each other using an axis (here the x-axis), then this alignment can be clarified (or solved) as a problem by first examining only point events that exist on the respective axis (here the x-axis) (Figures 1 and 20) (Einstein, 2009). This point exists with respect to the reference system K by means of the horizontal x and the system time t, with respect to K' these are x' and t', which can be derived if x and t are known (Einstein, 2009).

For more details on the comprehensible derivation of the Lorentz transformation, please refer to the original work by Albert Einstein (2009) (Reference section).

Equations (1) and (2) result for x' and t' with respect to a light beam moving laterally on the non-negative X' axis of the coordinate system K' at the speed of light c relative to K (Einstein, 2009).

(1) $x' = ax - bct$ (Space equation for K') (Einstein, 2009)

(2) $t' = at - \frac{bx}{c}$ (Time equation for K') (Einstein, 2009)

For a starting point of the light beam with respect to K' there is always x' = 0 and for each further movement of the light point in the same direction x' = 1, 2, 3 etc., according to equation (1), (3) and (4) now apply, which leads to (5), since bc/a corresponds to the system velocity v per unit time t or t' of K or K' (Einstein, 2009). The spatial equation for K' (1) can also be written like equation (3), analogous to equation (4) for the coordinates of the reference system K (Einstein, 2009).

(3) $x' = \frac{x}{a} + \frac{bc}{a}t'$ (Light body movement in K relative to K') (Einstein, 2009).

(4) $x = \frac{x'}{a} + \frac{bc}{a}t$ (Light body movement in K' relative to K) (Einstein, 2009).

If x' is now set in (3) or x = -1, -2, -3 etc. in (4), i.e. the opposite direction is selected for the light point movement on the non-positive X' or X axis, the system speed v per time unit t or t' of K and K' also remains constant in this model scenario, although the time in the form of the light beam moves back per event photo (snapshot) (Einstein, 2009).

In the author's opinion, the light propagation described by Albert Einstein (2009) is easiest to understand if readers imagine a movie of a moving body from two different locations; shots are taken of both locations per unit of time, i.e. the same movie is shot. This movie of a body from the universe can be viewed back and forth by the readers, for example a movie about the movement of the sun. To do this, one group of readers could record their matching movie version from the planet Mercury and the second group of readers would stay at home on Earth. Both groups would watch the sunrise, the first group would seem to see a larger sun compared to the second group of readers, yet all readers would observe the same body movement per unit of time in their respective frames of reference. As a result, the speed of the observed body can be determined exactly by both versions of a simultaneously recorded film; Albert Einstein (2009) uses the Lorentz transformation for this (with regard to a tiny body of light).

The velocity present in both systems can therefore be converted to b, which leads to equation (5) as explained above, in which equation (6) has already been integrated in a second step.

(5) $b = \frac{av}{c} = \frac{\frac{1}{\sqrt{1-\frac{v^2}{c^2}}}v}{c}$ (Einstein, 2009)

(6) $a = \frac{1}{\sqrt{1-\frac{v^2}{c^2}}}$ (Einstein, 2009)

Equation (6) can be determined on the basis of Albert Einstein's (2009) principle of relativity (Sections 6 and 25), i.e. derived by assessing a distance traveled by light propagation from K relative to K' using an immobile test rod; this path of light must be the same length from K' relative to K if the same immobile scale is used for this purpose. Albert Einstein (2009) therefore begins to make events (clearly) understandable from K on the X'-axis, he and his readers only need a photo of the event taken from K (snapshot); this means that they have to insert a freely selectable number (such as t = 0) for the system time K. If there is no system time t, they will get a photo of the event. If no system time t exists, you obtain ax for x' according to equation (1) (Einstein, 2009). Now Albert Einstein (2009) and his readers take another snapshot at time t = 1, on the basis of which they can interpret x' as 1, because they are looking at two events on the x' axis, each recorded (or measured) in the reference system K'. Both systems K' and K or the distances traveled in them therefore behave like (delta x) Δx to 1 by a (Einstein, 2009). However, if Albert Einstein (2009) and his readers take a photo of the event (instantaneous photo) from K' as before, starting at t' equal to 0, then with the help of equations (1) and (2) they now obtain equation (8) for x' by thinking away the system time t, taking equation (5) into account. Albert Einstein (2009) succeeds in thinking away the system time t by adding equations (1) and (2) and transforming them into (9).

These transformations are explained in an understandable way, i.e. in detail mathematically model-related to the universe with regard to the Lorentz transformation, because at this point Albert Einstein (2009) makes a mental leap that his readers will probably find (extremely) difficult to comprehend.

(7) $t = at' - \frac{bx'}{c} = t' = \frac{t}{a} + \frac{b}{ac}x'$ (Time equation for K) (Einstein, 2009)

(8) $t' = at - \frac{bx}{c} = t = \frac{t'}{a} + \frac{b}{ac}x$ (Time equation for K') (Einstein, 2009)

(9) $\frac{x'}{b} = ax - ct$ AND $\frac{ct'}{a} = ct - bx$ EQUAL $\frac{x'}{b} + \frac{ct'}{a} = ax - ct + ct - bx$ USE OF t‘ = 0 EQUAL $\frac{x'}{b} = ax - bx$ EQUAL $x' = [x(a-b)]b$ LOOKING BACK (5) EQUAL $x' = \left[x\left(a - \frac{av}{c}\right)\right]\frac{av}{c}$ EQUAL $x' = \left[\left(a - \frac{av}{c}\right)\frac{av}{c}\right]x$ EQUAL $x' = \left[a\left(a - a\frac{v}{c}\right)\frac{v}{c}\right]x$ EQUAL $x' = \left[a\left(1 - \frac{v}{c}\frac{v}{c}\right)\right]x$ EQUAL $x' = a\left(1 - \frac{v^2}{c^2}\right)x$ (Einstein, 2009)

From the transformation to (9), Albert Einstein's (2009) readers learn that on their further event photo, compared to the previous snapshot, a single distance can be seen between the two event points on the x-axis; relative to K, they obtain the distance according to equation (10).

(10) $\Delta x' = a\left(1 - \frac{v^2}{c^2}\right)$ (Einstein, 2009)

From the equation (10), a (6) can be derived, since Δx' = Δx or the right-hand side of (10) must be equal to 1 by a (11) as can be seen in the two event photos that Albert Einstein (2009) and his readers have mentally recorded.

(11) $\frac{1}{a} = a\left(1 - \frac{v^2}{c^2}\right)$ EQUAL $a^2 = \frac{1}{1-\frac{v^2}{c^2}}$ EQUAL $a = \frac{1}{\sqrt{1-\frac{v^2}{c^2}}}$ (Einstein, 2009)

Once the invariants a (6) and b (5) have been found in accordance with the previous assumptions, they can be reinserted into equations (1) and (2) accordingly (Einstein, 2009). The constant a (6) can be directly integrated into the constant b (5) for this purpose (Einstein, 2009). The velocity v characterizes the relative velocity of the two reference frames K and K' or from where the light beam is viewed in the space-time equations (Einstein, 2009).

In addition to the second (y' = y) and third (z' = z) equations, the Lorentz transformation is completely determined by inserting the constants a (6) and b (5) into the first (1) and fourth (2) equations (Einstein, 2009). Equations (12) and (13) now result for x' and t' with regard to a light beam that propagates laterally in a positive direction on the X' axis of the reference system K' with its velocity c relative to K (Einstein, 2009).

$$(12)\ x' = \frac{1}{\sqrt{1-\frac{v^2}{c^2}}}x - \frac{\frac{1}{\sqrt{1-\frac{v^2}{c^2}}}v}{c}ct = \frac{x}{\sqrt{1-\frac{v^2}{c^2}}} - \frac{vt}{\sqrt{1-\frac{v^2}{c^2}}} = \frac{x-vt}{\sqrt{1-\frac{v^2}{c^2}}}\ \text{(Einstein, 2009)}$$

$$(13)\ t' = \frac{1}{\sqrt{1-\frac{v^2}{c^2}}}t - \frac{\frac{\frac{1}{\sqrt{1-\frac{v^2}{c^2}}}v}{c}x}{c} = \frac{t}{\sqrt{1-\frac{v^2}{c^2}}} - \frac{\frac{\frac{v}{\sqrt{1-\frac{v^2}{c^2}}}}{1}x}{1} = \frac{t}{\sqrt{1-\frac{v^2}{c^2}}} - \frac{vx}{\sqrt{1-\frac{v^2}{c^2}}} = \frac{t(\sqrt{1-\frac{v^2}{c^2}})}{1-\frac{v^2}{c^2}} -$$

$$\frac{vx(\sqrt{1-\frac{v^2}{c^2}})}{1-\frac{v^2}{c^2}} = \frac{(\sqrt{1-\frac{v^2}{c^2}})(t-vx)}{1-\frac{v^2}{c^2}} = \frac{\left(1-\frac{v^2}{c^2}\right)(t-vx)}{\sqrt{1-\frac{v^2}{c^2}}} = \frac{t\left(1-\frac{v^2}{c^2}\right)-vx\left(1-\frac{v^2}{c^2}\right)}{\sqrt{1-\frac{v^2}{c^2}}} =$$

$$\frac{t\left(1-\frac{v^2}{c^2}\right)-\left(\frac{v}{1}-\frac{vvv}{c^2}\right)x\left(1-\frac{v^2}{c^2}\right)}{\sqrt{1-\frac{v^2}{c^2}}} = \frac{t\left(1-\frac{v^2}{c^2}\right)-\left(\frac{vvvv}{1c^2}-\frac{vvv}{c^2}\right)x\left(1-\frac{v^2}{c^2}\right)}{\sqrt{1-\frac{v^2}{c^2}}} = \frac{t-\frac{v}{c^2}x}{\sqrt{1-\frac{v^2}{c^2}}}\ \text{(Einstein, 2009)}$$

Using the four equations (12), (y' = y), (z' = z) and (13), this complete Lorentz transformation ensures that the law of the constant speed of light c is also observed in the (special) theory of relativity (Einstein, 2009). The law of light propagation applies in the (special) theory of relativity both in the respective reference system K and K' as well as for an arbitrarily selected direction of the light beam in the reference system K and K' (Einstein, 2009).

The Lorentz transformation still needs to be generalized (Figure 1) (Einstein, 2009). It is apparently not important for this that the coordinate axes of K and K' are aligned with each other, nor that a translation (movement) with a velocity v from K' to K has the same direction on the x-axis (Einstein, 2009). The general Lorentz transformation consists of itself in particular and of a space transformation, which corresponds to the exchange of a reference system at right angles with another, new coordinate system with angularly deviating axes (Einstein, 2009).

The result is a spherical geometry according to the general Lorentz transformation, which mathematically expresses x, y, z, t as linear properties of the same type with regard to x', y', z', t' (Einstein, 2009). The relationship (14) therefore applies (Einstein, 2009). This general Lorentz transformation contains σ equal to 1 on its left-hand side (14) with regard to all events located on the x-axis, since the sequence (14) can be verified by eliminating (thinking away) y and z as well as y' and z' with regard to x or x' (12) and t or t' (13) (Einstein, 2009).

(14) $\sigma(x^2 + y^2 + z^2 - c^2t^2) = x^2 + y^2 + z^2 - c^2t^2 = x'^2 + y'^2 + z'^2 - c'^2t'^2 = (x'\sqrt{1 - \frac{v^2}{c^2}} + vt)^2 + y^2 + z^2 - c^2(t'\sqrt{1 - \frac{v^2}{c^2}} + \frac{v}{c^2}x)^2 = (\frac{x-vt}{\sqrt{1-\frac{v^2}{c^2}}})^2 + y'^2 + z'^2 - c'^2(\frac{t-\frac{v}{c^2}x}{\sqrt{1-\frac{v^2}{c^2}}})^2$ (Einstein, 2009)

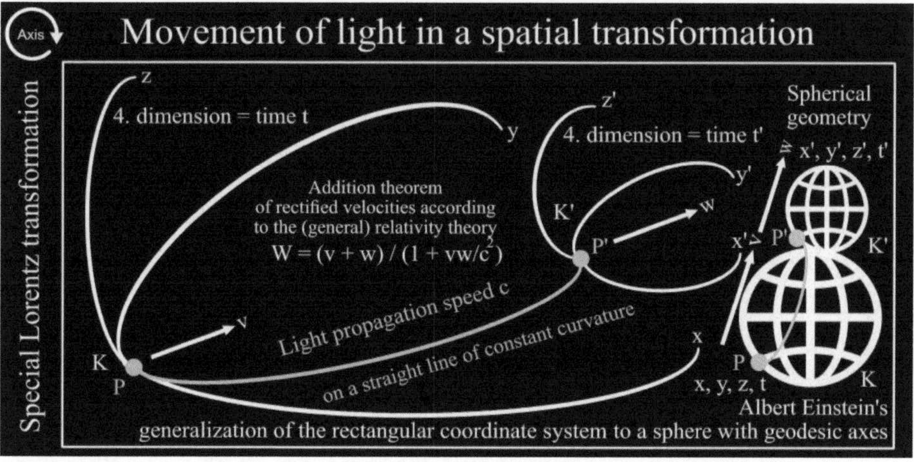

Figure 1. Spherical geometry in Albert Einstein's (general) theory of relativity.

2 The Minkowski universe with four dimensions

This general transformation based on Lorentz can also be described better, i.e. more accurately (more precisely), by generally implementing an idea of the time c_it_i for K or $c'_it'_i$ for K' in the four-dimensional model of the universe, instead of continuing to accept the first variable t for K and the second variable t' for K' for the time in the model system of the universe on the fourth dimension in general (Einstein, 2009). According to Minkowski, equations 15 to 18 result for these four dimensions in a more general notation according to (theoretical) physics (Figure 2) (Einstein, 2009).

(15) $x_i = x = x' = x'_i$ (Einstein, 2009)

(16) $x_i = y = y' = x'_i$ (Einstein, 2009)

(17) $x_i = z = z' = x'_i$ (Einstein, 2009)

(18) $x_i = c_it_i = c'_it'_i = x'_i$ (Einstein, 2009)

Equations 15 to 18 result in a sequence (19), Albert Einstein (2009) and his readers set i = 1, 2, 3 and 4, starting from K relative to the reference system K', which is given or conditioned by the general transformation based on Lorentz.

(19) $x_i^2 + x_i^2 + x_i^2 + x_i^2 = x'^2_i + x'^2_i + x'^2_i + x'^2_i$ (Einstein, 2009)

Due to this described selection (determination) of x_i point coordinates, equation (14) is taken into account in the sequence (19), even if a movement (change) of time is now assumed for the static system time in K and K', i.e. a dynamic system time in K and K' (Einstein, 2009).

This is (also) generally evident in the sequence (19), in that an imaginary x_i-time coordinate (for the system time in the model of the universe) is contained in the transformation equation (14) in exactly the same way as all three other x_i space coordinates (Einstein, 2009). This is the reason why general laws of nature concerning the system time x_i as well as the system space x_i as four point coordinates are taken into account in the [general and thus also in the special (as well as in the general-special)] theory of relativity (Einstein, 2009).

Minkowski selected (determined) for each x_i as a universe point (or an event coordinate) that i = 1, 2, 3 and 4, which makes it possible to represent the universe (continuum) with four dimensions (as a four-dimensional model) (Einstein, 2009). So if in a space with three dimensions its physics represents an event (time-dependent caused or happened by objects in the universe such as the temporal meeting of two bodies), then in a space with four dimensions its physics (again) represents existence (the time-independent connection of all bodies in the universe such as that of humans on Earth in the solar system in the Milky Way, etc.); provided a universe (continuum) with the corresponding number of dimensions really exists (Einstein, 2009).

For a better understanding for the readers of this book based on Albert Einstein (2009) concerning the difference between event and existence: Because a human being collides with the universe around him as an object due to his x_i point coordinates (event occurring at all times) and therefore simultaneously represents a body in the universe, the same (general) laws of nature apply to him considered as a body as in the universe (continuum). This is because, according to Albert Einstein's (general) theory of relativity (2009), the human being is rather to be understood as a body existing (enclosed) in the present that is connected to the universe (continuum) (existence). The question of the existence of man (event) should be explained by the fact that according to this theory of relativity by Albert Einstein (2009), system time is generally accepted (lived) as the fourth dimension, whereby the existence of man (event) should be like the timeless (infinite) universe (continuum) in which this existence of man (event) takes place (existence).

A (non-Euclidean) universe with four dimensions (reference system K) has the (mathematically) rooted analogy according to the geometry on which this study is based with respect to a universe with three dimensions (reference system K'), which adheres to Euclid's geometry, whereby both types of universes appear approximately the same despite the different number of dimensions (Einstein, 2009). If, for example, the more modern Cartesian reference system (coordinate system) with x'_i events and for i = 1, 2 and 3 is introduced for a universe with three dimensions K' as the simultaneous starting point of the expansion of the three-dimensional universe K' and four-dimensional universe K, then x'_i of K' mean linearly similar properties with respect to x_i of K, since these (space) properties comply with condition (20) in the same way as if additional (time) properties were included as in condition (19) (Einstein, 2009).

(20) $x_i^2 + x_i^2 + x_i^2 = x'^2_i + x'^2_i + x'^2_i$ (Einstein, 2009)

A relationship to (19) is completed by this model superposition (model analogy) between the reference systems (universes) K and K' (20) (Einstein, 2009). A universe (continuum) according to Minkowski is generally to be understood according to its laws of nature in the same way as a space with four dimensions in which Euclid's geometry applies [due to the imagined (fantasized) coordinate for the system time]; a universe (continuum) according to Minkowski with (or from) four dimensions as a reference system (coordinate system) has a rotation in the (general) theory of relativity due to the transformation according to Lorentz (Einstein, 2009).

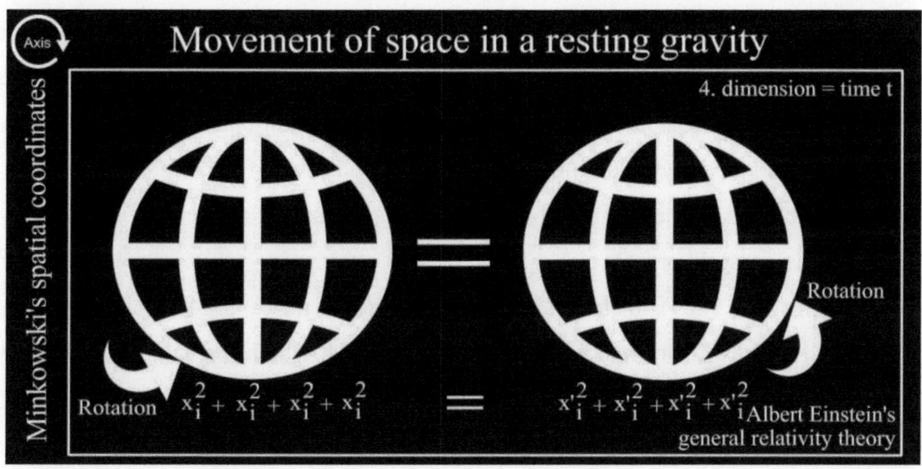

Figure 2. Motion of space in Albert Einstein's (general) theory of relativity.

3 On the general relativity theory confirmed by the experience theory

Experience theory is not a flow of deductive thoughts, but a flow of inductive thoughts that requires a continuous template-like, knowledge-based model perspective, which (in turn) enables a (systematized) process of the emergence (of knowledge) according to experience theory (Einstein, 2009). These insights (assumptions) represent (only) condensed thoughts (summaries) on the basis of large sums of individual insights (e.g. given in specialist articles) which are (always) understood as (shared, i.e. common) experiential propositions which undergo a comparison to determine experience-based (established) more general propositions (theories) (Einstein, 2009). From this model perspective, the progress of all sciences acts like a two-dimensional knowledge matrix (cataloging model), (i.e.) like the model result of pure empirical research (Einstein, 2009).

However, the previous model perspective (way of thinking) in no way invalidates a physical-real experienceability due to the induction thought flow (an experience flow due to reference bodies, such as specialized articles, to generate a flat knowledge matrix, the empirical model science) (Einstein, 2009). This model perspective does not include, for example, a central task such as inspiration (idea generation) and the flow of deductive thought, which is necessary for the progress of a single true (precise) form of science (Einstein, 2009). For example, at the moment when this form of science has surpassed even the simplest model maturity of knowledge, the cause of experience-based (theory-thought-based) developments of the scientific model is by no means possible only by means of a purely cataloguing (systematizing, ordering) task (Einstein, 2009). Instead, each scientist should make targeted progress, (only) inspired by physically real experience situations, in which the scientists build (and visualize) a (shared, i.e. common) systematized thought structure (model matrix), which according to logic corresponds to a (largely) low number of basic ideas (principles, propositions, axioms) (Einstein, 2009). Albert Einstein (2009) describes an (individually) systematized thought structure (model matrix) as an overall model assumption (all model hypotheses, meta-thought model) (or individually conceived theory components of the overall model of science). Each overall model (meta-mind model) of science draws its claim to existence from a higher (and growing) number of individual experience situations, which are all connected with each other in the overall model, which is why there is no falsity in this overall model (Einstein, 2009).

If the overall model (meta-mind model) of science is considered as a collection of individual experience situations with high complexity due to numerous connections between the individual experiences, different model hypotheses (model assumptions, model thoughts) can arise, which, when put together, nevertheless have no overlap and are therefore not really the same (Einstein, 2009). Precisely because this overlap of model assumptions (model theories) can appear extremely large due to their consequences achievable through experience theory, it is (only extremely) difficult (not only for scientists) to comprehend these conceivable (experienceable) consequences according to (collected individually) experienced experiential situations, with regard to which the overall model (meta-thought model) of science can be differentiated in terms of content, namely on the basis of its individual theory-based (thought-based) model components, which are, however, similar in content (Einstein, 2009).

A corresponding overlap between two fundamentally different model theories in the overall model (meta-thought model) of a science can be found, for example, in the (still two-dimensional) scientific field of biology, whose theories contained in the model are of more general importance for living beings such as humans (Einstein, 2009). In biology, a distinction can be made between evolutionary theory and Darwinian theory, both of which have in common that they provide hypothesis-based justifications for how the evolution of living beings (in this ever faster expanding universe) takes place (Einstein, 2009). In the first theory, this is based on the assumption that once acquired characteristics (such as knowledge) are inherited, and in the latter theory, this is based on a differentiated assumption, namely that a dispute about (four-dimensional) existence is won by selecting the breeding (of only strong living beings) (Einstein, 2009).

In Albert Einstein's (general) theory of relativity in comparison with Isaac Newton's theory of gravitation, there is the described scenario of a high overlap of consequences (between a number of theories, i.e. between two theories that are both (mentally) recognized in the overall model (meta-mind model) of physics) (Einstein, 2009).

The present overlap between Einstein's (general) theory of relativity and Newton's theory of gravitation is very large, so that until today only a small number of experiential (conceivable) consequences of Einstein's (general) theory of relativity could be discovered according to the theory of experience (inductively), with regard to whose consequences due to the relativity of gravitation of reference bodies an outdated (one-sided) physics model by no means led – although there is a momentous differentiability between the basic principles of both theories (Einstein, 2009). In his section 3 (on the theory of experience), Albert Einstein (2009) would like to examine the consequences to be emphasized (i.e. suitable for differentiation) one last time (inductively) (on the basis of two theories of gravitation contained in the overall physics model), as well as discussing all individual experiences collected to date in this regard (therefore only) very briefly.

3.1 The perihelion point of the planet Mercury on its elliptical motion

A planet flying alone, such as Mercury, orbiting a star, such as the Sun, should form an elliptical motion around this star (or more precisely around a split center of mass between the star and the planet) according to the classical theory of motion and the classical theory of gravity (Figure 3) (Einstein, 2009). The star (or the divided gravitational center of mass) is located within the elliptical orbit in a collecting point (focal point), so that the distance between the star and the planet increases from a minimum value to a maximum value and then shrinks back to the minimum value in the course of the planetary orbital year (Einstein, 2009). If a different law is generally inserted into the equation instead of the classical law of gravitation, then it is generally stated that this orbital motion should proceed unchanged according to this theory, so that the distance between the star and the planet alternately increases and decreases; however, the angle between the star and the planet given by a (uniform) straight line would differ by 360 degrees [from the perihelion point (near the star) to the perihelion point (far from the star)] at such a time period (Einstein, 2009). This straight line of his orbital ellipse would therefore represent an open one (due to the angular deviation), which within the course of time forms a ring-like area of the orbital surface (between a circular year with respect to the minimum planetary distance and the circular year with respect to the maximum planetary distance) (Einstein, 2009).

According to the general theory of relativity, which is known to differ slightly from the classical theory, there will now also be a similar slight difference with respect to Kepler-Newtonian orbital mechanics, so that the angle given by a radius straight line from the star to the planet between the perihelion point and an angle below it with respect to an entire orbital revolution angle (which in physics means from the angle 2π as the usual absolute angle measure) differs by the expression (21) (Einstein, 2009). (Where a represents the larger semi-axis of the ellipse, e stands for the eccentricity of the ellipse, c indicates the speed of movement of light, and T the orbital period) (Einstein, 2009).

(21) $24\pi^3a^2 / T^2c^2(1 - e^2)$ (Einstein, 2009)

In general, this can also be formulated accordingly: according to the general theory of relativity, the larger elliptical axis rotates analogously to a linear motion around a star (Einstein, 2009). According to mechanics, the rotation of the planet Mercury must be 43 arc seconds every 100 years, but with regard to the other planets around our star, the sun, it must be extremely small, so that it must not be possible to determine (state) the rotation (Einstein, 2009).

In reality, astronomers found that classical mechanics is too imprecise to calculate the observed orbital motion of Mercury with an accuracy achievable by today's observational methods (Einstein, 2009). Taking into account all the variable influencing forces that all the other planets have on Mercury, it becomes apparent (Lever-Rier 1859 and Newcomb 1895) that an unfounded perihelion motion of the Mercury orbit line continued to exist, which in no way deviated noticeably from the other planetary orbits by +43 arc seconds every 100 years (Einstein, 2009). This deviation of the empirical result, which agrees with the result of the general theory of relativity, amounts to a small number of arc seconds (Einstein, 2009).

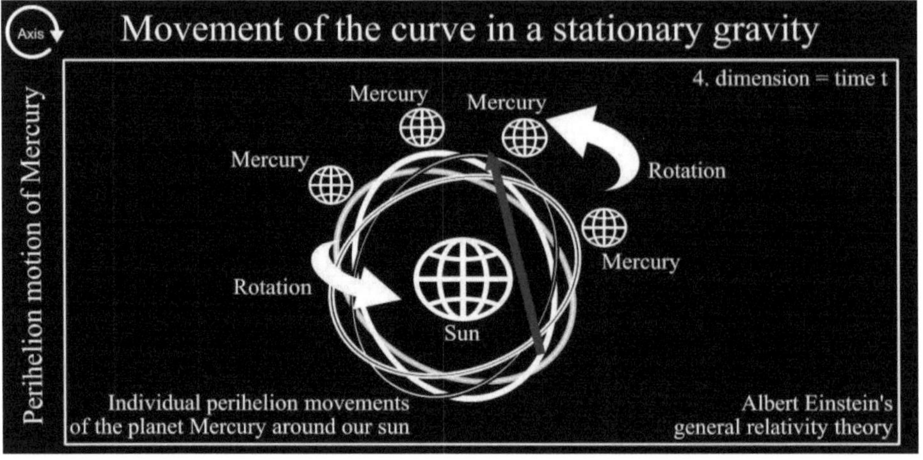

Figure 3. Movement of the orbital curve in Albert Einstein's (general) theory of relativity.

3.2 The redirection of light movement caused by the gravitational field

Section 10 describes that, according to the general theory of relativity, a ray of a light body must be bent by means of a gravitational field in a manner similar to the bending that a trajectory line must experience from a reference body (mass point) accelerated by a gravitational field (Einstein, 2009). According to Albert Einstein's general theory of relativity (2009), a ray of light passing a space-time body must (first) be redirected towards this body; the angle of redirection α must be the same for this ray of light passing this celestial body at a distance of Ω star radii,

(22) $\alpha = 1{,}7$ Seconds / Δ (Einstein, 2009)

make out (Einstein, 2009). Albert Einstein (2009) adds that, according to the theory, 50 percent of this detour is caused by the (classical) gravitational field of the star, our sun, and 50 percent by the change in the geometry ("bending") of space caused or (self-)created by this star.

This result made it possible to verify the experiment by means of photographic snapshots of the sun during a total eclipse (Einstein, 2009). This solar eclipse should only be awaited because at any other time the atmospheric radiation illuminated by starlight shines correspondingly brightly, so that all suns close to the stars (fixed stars) cannot be seen (Einstein, 2009).

The expected picture can be easily recognized from Figure 4 (Einstein, 2009). If no star existed, then a practically (almost) infinitely distant sun would generally be observed in the direction R_1 (Einstein, 2009). However, due to the redirection by the star, this fixed star is generally observed in the direction R_2, i.e. at a slightly greater distance from the center of the star than corresponds to physical reality (Einstein, 2009); the fixed star seen should therefore be at a different location further to the left instead of to the right due to the redirection of light.

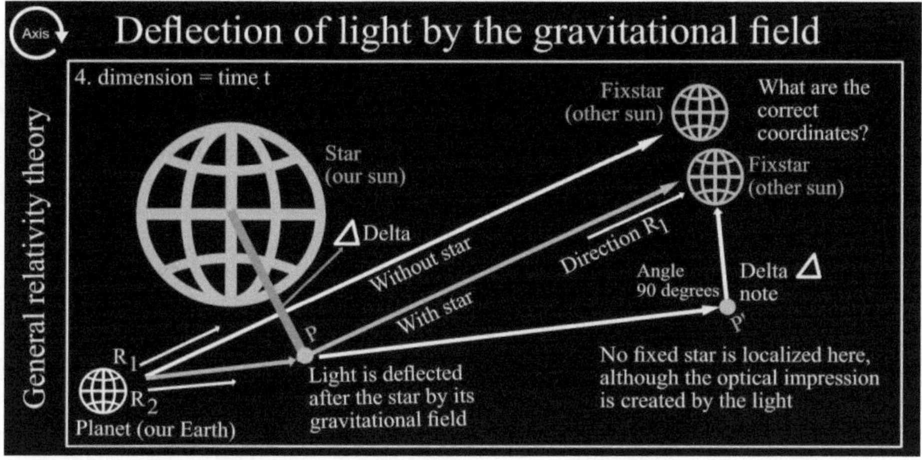

Figure 4. The redirection of light seen through the gravitational field of a star, i.e. observed from R_1.

In practice, this check is carried out using the following method (Einstein, 2009). All fixed stars in the space-time area of the star, in this case our sun, are photographed during a (total) eclipse (Einstein, 2009). In addition, a second photo snapshot of the same fixed stars is taken as soon as the star is at a different location in the space-time range (i.e. several time periods before or after) (Einstein, 2009). Accordingly, the solar photos taken during a stellar eclipse should not have moved radially inwards (but outwards from the center of the star) by an amount equal to the (previous) angle α with respect to the comparison snapshot (Einstein, 2009).

Albert Einstein (2009) and his readers can be grateful to the Astronomical Royal Society for verifying this groundbreaking result. Not to be deterred by the First World War and the mental strain it caused, the Society sent three of its most important astronomers (Eddington, Crommelin and Davidson) and provided two expeditions with equipment to record the eclipse on May 29, 1919 in Sobral (Brazil) and on the island of Principe (West Africa) (Einstein, 2009). The expected relative deviations of the star dimness images compared to the star brightness images were only a few hundredths of a millimeter (Einstein, 2009). The requirements for the accuracy of the snapshots and their physical dimensions were therefore high (Einstein, 2009).

The result of this measurement proved the general theory of relativity in an absolutely satisfactory way (Einstein, 2009). The components in the right angle of the seen and calculated deflections (deflections) of the light rays of the fixed stars (measured in arc seconds) are recorded in the next table (Einstein, 2009):

Star number	**1. Coordinate**		**2. Coordinate**	
	observed	**calculated**	**observed**	**calculated**
2	+ 0.95	+ 0.85	– 0.27	– 0.09
3	– 0.20	– 0.12	+ 1.00	+ 0.87
4	– 0.11	– 0.10	+ 0.83	+ 0.74
5	– 0.29	– 0.31	– 0.46	– 0.43
6	– 0.10	– 0.04	+ 0.57	+ 0.40
10	– 0.08	+ 0.09	+ 0.35	+ 0.32
11	– 0.19	– 0.22	+ 0.16	+ 0.02

Table 1. Observed and calculated star deviations measured in arc seconds (Einstein, 2009, p. 86).

3.3 The red change of spectral lines

Section 11 shows that within a reference system K', which rotates against a Galilean coordinate system K, the directional speed (of the movement of time or) of hand rotations (hand movements) of clock machines of the same type placed at rest is not independent of their position (Figure 5) (Einstein, 2009). Albert Einstein (2009) and his readers would like to think through this relationship (dependence) in terms of quantity. The clock hand machine, which is placed at a distance r from the center of the disk, has the directional velocity relative to K

(23) $v = wr$ (Einstein, 2009),

if w is the rotational speed of the circular plane (K') with respect to K (Einstein, 2009). If v_0 is the number of hand movements on the clock face per time period (uniform time coordinate) (running speed) relative to K when the clock face is fixed or not moving, then the hand running speed v of the clock face planted with a directional speed v relative to K and fixed (at rest) relative to the circular plane is equal to the speed of the clock face according to section 32

(24) $v = v_0\sqrt{(1 - v^2/c^2)}$ (Einstein, 2009),

or supplemented by sufficient precision

(25) $v = v_0\sqrt{(1 - 1/2 \times v^2/c^2)}$ (Einstein, 2009),

or also to be equated with

(26) $v = v_0\sqrt{(1 - w^2r^2/2c^2)}$ (Einstein, 2009).

If the deviation of the power potential of the centrifugal interaction between the position of the watch dial and the center point of the circle, i.e. the non-positive absorbed power, which is generally added in the opposite direction to the centrifugal interaction of the uniform mass point so that it can be conveyed (transported) from the watch resting point on the moving circular plane to the center point, is generally calculated using $+\Omega$, then

(27) $+\Omega = - w^2r^2/2$ (Einstein, 2009),

then generally

(28) $v = v_0\sqrt{(1 + \Omega/c^2)}$ (Einstein, 2009).

From this it is generally first recognized that two clock faces of the same composition (nature) with a different distance from the centre point of the circular plane move at different speeds (not the course is meant as that of the passage of time, but only the movement of the clock hands), this result is also valid from a standing point of the eye witness (observer) rotating with this circular plane (Einstein, 2009).

Because now – starting from our circular plane – a gravitational field exists, which has the action potential Ω, then the generated result must only be valid with regard to gravitational fields (Einstein, 2009). Because Albert Einstein (2009) and his readers can additionally regard spectral lines as an emitting particle (quantum) as the time (clock hand machine), then the axiom is valid:

Each particle can absorb or emit a frequency number that is not independent of the effective potential of a gravitational field within which the quantum moves (Einstein, 2009).

The oscillation number (frequency number) of any particle moving associated with a body surface must be slightly less than an oscillation of a particle of the same mass element moving within the body-free universe space (or associated with a smaller body surface within our world) (Einstein, 2009). Because Ω is analogous to –KM/r, where K stands for the classical gravitational field constant, M for the mass, r for the radius of the world body, then a red change in a spectral line produced on a stellar ellipsoid surface compared to the spectral line produced on the Earth's ellipsoid surface should be equal to the value

(29) $(v - v_0)/v_0 = -KM/c^2r$ (Einstein, 2009)

be observed (Einstein, 2009).

In the case of our sun, the red change to be assumed comprises approximately two times one millionth of the wavelength of light (Einstein, 2009). The mass M and the radius r are generally not clearly given for the fixed stars, so no true (plausible) calculation is conceivable for these stars (Einstein, 2009).

Whether the redshift effect (of the gravitational fields) really exists today corresponds to a closed question, the answer to which, in retrospect, would have to be worked out by all astronomers with considerable motivation (Einstein, 2009). In the case of our solar body, the existence of this effect appears difficult to observe due to its smallness (Einstein, 2009). Researchers such as Grebe, Bachem, Evershed, Schwarzschild and Perot all consider the existence of the effect to be firmly predetermined (by nature) on the basis of their own measurement observations (and in the case of Grebe and Bachem also by Evershed and Schwarzschild), whereas other researchers, above all Julius and John, are of the opposite opinion due to their measurement observations or are by no means certain of the possibility of proving the empirically available data observation material to date (Einstein, 2009).

Within the statistical analyses with regard to the fixed stars, average straight line changes in the direction of the non-short-wave spectral dimension always appear to exist (Einstein, 2009). However, the evaluation of the (subjective) observation material available to date still does not allow any convincing assumption to be made as to whether this change is actually due to a gravitational field effect (Einstein, 2009). A summary of the measurement observations, including an extensive debate based on the point of view of the question that interests Albert Einstein (2009, p. 89) and his readers, is generally contained in the paper by Freundlich E. entitled "Prüfung der allgemeinen Relativitätstheorie" published in "Die Naturwissenschaften 1919, H. 35, p. 520" by "Verlag [...] Springer, Berlin".

Obviously, the following calendar years must provide a convincing decision assumption (Einstein, 2009). Since a red change of the spectral line exists by means of the gravitational action potential, Albert Einstein's (2009) general theory of relativity is consistent. At the same time, once this knowledge is understood as a consequence originally based on the gravitational action potential, the understanding of the line change must bring significant thoughts regarding the body masses in our universe (Einstein, 2009).

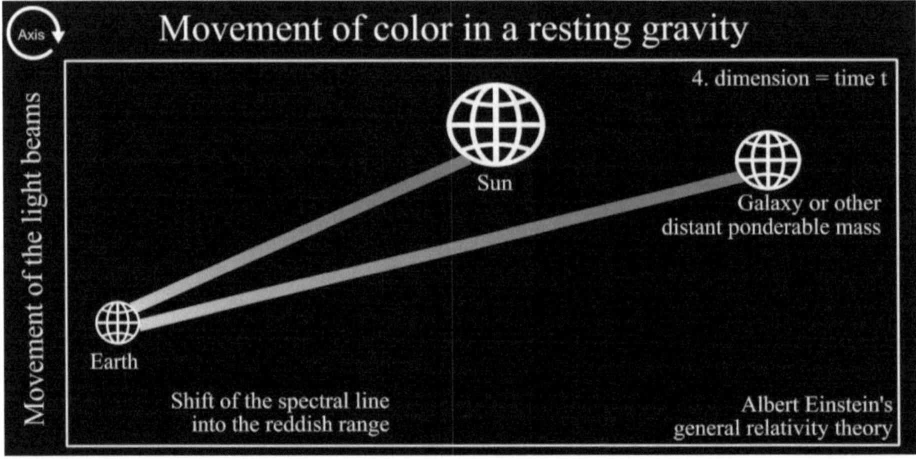

Figure 5. The coloration of light as seen through the gravitational field of a star, i.e. observed from Earth.

4 The spatial structure in connection with Albert Einstein's general relativity theory

Since the publication of the very first edition of his book (1916), the experience of Alberst Einstein (2009) and his readers about the spatial structure within the whole (cosmos problem) has undergone a significant evolution, which can also be noticed (or learned or experienced) within a (universally) understandable representation of this subject (Figure 6).

Albert Einstein (2009) had assigned two assumptions (theories) as foundations to his previous (initial) thoughts on the subject of spatial structure:

1. In the spatial structure as a whole, there is a non-zero average density of matter, which is the same at every location (Einstein, 2009).

2. The propagation (or the "radius line") of the spatial structure as a whole is not dependent on time (Einstein, 2009).

For Albert Einstein (2009), these two theories (constellations, derivations) were linked according to his general theory of relativity, but only in the case where an assumed term was generally added to the gravitational field equations, which his theory of relativity did not require in terms of content, and also did not act from the conceivable theoretical point as given by nature ("term about cosmology in the gravitational field equations").

His second assumption seemed necessary to Albert Einstein (2009) at this point in time, because he noted in retrospect (an expectation thought), (in the form) that could generally be dragged (slipped) into infinite discussion hypotheses (thought speculations), should there be a general deviation from this second hypothesis.

As early as the 1920s, however, the Russian-born mathematician Friedman found that a diversified (deviating, different) hypothesis existed that was more natural from the completely theoretical point of view (Einstein, 2009). Friedman understood, for example, that it could not be ruled out that the first assumption could be maintained without inserting the rather unnatural term within the gravitational field equations as soon as there was a general decision to reject the second assumption (Einstein, 2009). Albert Einstein's (2009) initial gravitational field equations, for example, allow a solution system (reference system) within which a "radius line of the continuum (universe)" is not independent with respect to time (space expansion). According to this point of view, it is generally possible to assume, in agreement with Friedman, that Albert Einstein's (2009) general theory of relativity may require (but does not require) an expansion of space.

Barely a few ellipsoidal orbit years later, Edwin Powell Hubble used his spectral analyses of extra-galactic cloud nebulae ("milky cloud streets") to illustrate that the spectral lines sent out by the clouds exhibit a uniformly increasing red change with the distance of the clouds (Einstein, 2009). According to Albert Einstein (2009) and the experience of his readers today, the change in red is to be understood in accordance with the Doppler principle merely as a growth movement of the (observed) stellar body system (reference system, coordinate system) within the whole – as this requires the analysis of the gravitational field equations according to Friedman (as a consequence). This finding (discovery, observation) by Hubble is therefore to be understood as an evaluation of the (general) theory of relativity (Einstein, 2009).

For Albert Einstein (2009), however, there is a serious incongruence, i.e. an insufficient agreement (coincidence) as a restriction (limitation) of Hubble's observation (discovery, detection). An (intellectually unassailable) interpretation (interpretation assumption) of the straight line changes assigned to the galaxies discovered by Hubble as a spatial growth leads to an expansion start that "only" began about 10 to the power of 9 calendar years ago, although (theoretical) astronomical physics considers this to be possible with a high degree of certainty that the formation of all stellar bodies and stellar body systems required significantly longer periods of time (Einstein, 2009). It is still not entirely clear how these missing coincidences (incongruences) can be compensated for (solved with equations) (Einstein, 2009). This is because there is still a lack of experience in the astronomical and physical sense of understanding and observing stars and star systems outside our solar system.

Albert Einstein (2009) also wants his readers to learn that an assumption about the expansion of space in connection with all (empirical) observations of measurement data collected by astronomy does not allow any decision theory regarding an infinite or finite space (on three dimensions), although the initial mathematical assumption of space did not allow any openness (infinity) of a space as a result.

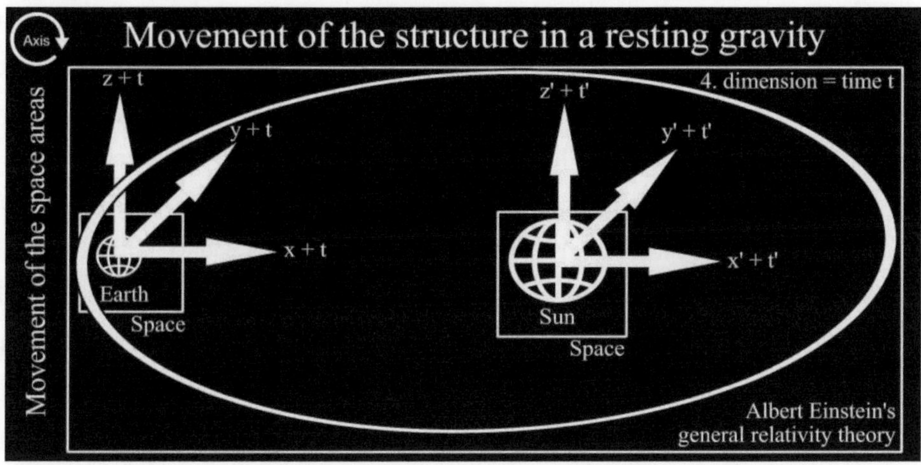

Figure 6. The spatial structure seen through the gravitational field of a star, i.e. observed from Earth.

5 Space theory and its relativity in general

It is characteristic of classical physics that, in addition to matter, it assigns a non-dependent real existence to both space and time (Einstein, 2009). This is due to the fact that the definition of acceleration originates from the classical theory of motion (Einstein, 2009). In classical theory, however, acceleration merely expresses "directional acceleration in space" (Einstein, 2009). Classical space must therefore be "immobile", or at the very least be regarded as "not accelerated", so that in general the directional acceleration, which originates from the theory of motion, can be regarded as a value in a physical sense (Einstein, 2009). The same applies to time, which is of course also included in the definition of acceleration (Einstein, 2009). Even Newton and his critics at the time found it illogical that a physical reality could be assigned to space and its state of motion in general; however, at that time there was no other possibility if a meaningful understanding was to be assigned to motion in general (Einstein, 2009).

It is already a strange theory that in general a physical reality can actually be assigned to a space, especially an unfilled space (Einstein, 2009). Philosophers have repeatedly resisted such an assumption since the beginning of science (Einstein, 2009). Descartes reasoned in approximately the same way (Einstein, 2009): Space appears to be equivalent with respect to expansion, but expansion appears to be dependent with respect to bodies. Accordingly, there is no space without bodies, which means no unfilled space (Einstein, 2009). This is the main limitation of such a conclusion (Einstein, 2009): It is correct, of course, that the definition expansion has its origin in the experience that Albert Einstein (2009) and his readers had with regard to (solid) bodies during storage (touching). However, no general conclusion can be drawn from this that the definition of expansion is unfounded within scenarios that should not give rise to the formation of this definition (Einstein, 2009). Such a dimensioning with regard to definitions also makes sense in a roundabout way by means of its contribution to an understanding of empirical results (Einstein, 2009). A hypothesis that expansion is linked to bodies does not appear to be substantiated (Einstein, 2009). However, Albert Einstein (2009) and his readers have to understand later on that Descartes' opinion is indirectly proven by the general theory of relativity, which is to be noted with a grain of truth. The reason why Descartes thought this seemingly strange opinion must probably have been the emotion that, in general, no physical reality can be attributed to an (only) "immediately perceptible" object such as a space with a missing, urgent requirement (Einstein, 2009).

A mental origin of this definition of space, or the need for it, appears to be much more unconscious, as may be true because of the peculiarities of human psychology (Einstein, 2009). All earlier geometry experts used psychological objects (lines, points, surfaces), but none did so in terms of space as a space, as later investigative geometry mathematicians did (Einstein, 2009). However, the definition of space appears to have been inspired by many a simple experience (Einstein, 2009). In general, a box could have been set aside (Einstein, 2009). Objects can generally be placed in it in any systematization so that the box fills up (Einstein, 2009). The existence of these (possible) systematizations resembles a function of the objective body box, a thing that exists together with a box, a space "encompassed by a box" (Einstein, 2009). This is a thing that exists differently with respect to different boxes, a thing that is completely conceivable as given by nature with no dependence on the box, regardless of whether there are actually objects inside this box or none are placed inside it (Einstein, 2009). If there are no objects inside the box, then its space appears "unfilled" (or "open") (Einstein, 2009).

Albert Einstein's (2009) and his readers' definition of space appears to be linked to the box. For Albert Einstein (2009) and his readers, however, it becomes apparent that all systematization possibilities describing the box space cannot be dependent on how wide the box boundaries are. Is it not generally possible to reduce this width to zero, but without accepting the (intended) "space" (of the box) as a loss (Einstein, 2009)? The simplicity of such a thought boundary process seems understandable, and now there exists for human psychology a space with a missing box, the independent object, which nevertheless seems extremely unreal if the origin of this definition is generally neglected (Einstein, 2009). It is generally understood that Descartes did not like to look at space, independently in terms of representational bodies, as an object that should exist despite the absence of matter (Einstein, 2009). Kant's approach of trying to eliminate this imposition by denying the objecthood of space should therefore not actually be understood as a theory (Einstein, 2009). The systematization possibilities, geometrically given due to the storage space of the box, do not appear to be subjectively designed in the same sense as a box per se and all objects that can be placed in this box (Einstein, 2009). (However, the lack of matter in no way disturbs Descartes from regarding this space as a fundamental definition within the geometrical analysis he derived (Einstein, 2009). The comment regarding a vacuum in a "mercury barometer" certainly renders the remaining dualists (Cartesians) (argumentatively) defenceless (Einstein, 2009). However, it does not seem to be deniable that even at this simplest stage of development the definition of space or our space, recognized as an independent real object, has a little incompleteness (Einstein, 2009).

The ways in which bodies can be placed in spaces (boxes) are similar to the object according to Euclid's geometry in three dimensions, which, due to its axiomatic design, can quickly give the false (illusionary) impression that it can be linked to tangible states (Einstein, 2009).

If, according to the method described above, after an experience of "filling" a box, the definition of space now appears to be designed, then this idea initially represents a limited space (Einstein, 2009). However, the limitation becomes unimportant, as it generally appears that a larger dimensioned box can always be introduced, which includes a smaller dimensioned box (Einstein, 2009). A space thus becomes a material unlimited (or infinite) (Einstein, 2009).

Albert Einstein (2009) does not want to talk at this point about the fact that the understanding of the three dimensions and the "Euclidianity" (equality) of space are based on (relatively simple) experiences, instead he would like to illustrate the task of defining space within the improvement of the way of thinking in physics with different points of thought.

If a smaller dimensioned box b lies relatively at rest inside the empty space of a larger dimensioned box B, then the empty space of b resembles a component of the empty space of B, and the same "spatial construct" encompassing these two boxes counts as these two boxes (Einstein, 2009). However, it seems to be more difficult to understand if b is moved with respect to B (Einstein, 2009). The general tendency is to understand that b always comprises the same space, but it is a non-constant (variable) spatial component of space B (Einstein, 2009). In general, it seems necessary to attribute to all boxes their special (unboundedly assumed) space and to think that the two space constructs exist one against the other not resting (Einstein, 2009).

Only after a general awareness of this confusion (chaos) (of spaces) has been developed does our space no longer seem analogous to a confined container (medium) within which the objective bodies whirl around (Einstein, 2009). Now, however, Albert Einstein (2009) and his readers in general must understand that there is no finite number of spaces, all of which exist propagated against all spaces. The construct of space analogous to a non-dependent objectivity existing with respect to bodies is already part of the popular understanding, but a scientific understanding is the flash of thought regarding the existence of an infinite number of oppositely non-resting spaces (Einstein, 2009). This last idea certainly appears to be necessary from an intellectual point of view, but for a long time it had no major role as an approach within scientific thinking (Einstein, 2009).

But what is at stake in the conceptual origin of the definition of time (Einstein, 2009)? This construct can be reliably related to the reality of "human memory", and to the difference between (present) sense experience and (imagined) experience as a conception of meaning (Einstein, 2009). In fact, it must be questioned here whether this difference between sense experience and experience (or pure fantasy) represents a thing that is directly objective in thought (Einstein, 2009). Everyone has experienced that they doubted whether they were experiencing things in a meaningful way or only dreaming about them (Einstein, 2009). It can be assumed that the described difference arises as a thought after the assigning meaning (action) (Einstein, 2009).

We assign an experience of meaning to our "experience", which is understood as "before" compared to the "present experiences of meaning" (Einstein, 2009). This is a creative systematization principle with regard to (psychological) sense experiences, the applicability of this principle leads to a justification of the non-objective time construct, i.e. this understanding of time, which is similar to the systematization of sense experiences of individuals (Einstein, 2009).

Non-subjectivization of the time construct (Einstein, 2009). Thought conception (Einstein, 2009). Person A ("Albert Einstein") sees a "flash" as a sensory experience

(Einstein, 2009). Albert Einstein (2009) also experiences (learns) the same behavior of person B ("I"), who thinks the behavior of Erik Kolek with his sense experience "lightning seen" in a relation (equation). This is the reason why Albert Einstein (2009) attributes (mentally) his (or our common) sense experience "lightning seen" to me. Albert Einstein (2009) comes to the understanding that other people are also involved in this "lightning seen". The "lightning seen" now does not appear as a purely individual sense experience in the thoughts, but instead as a sense experience (or now merely as a "possible sense experience") of other people (Einstein, 2009). Accordingly, the understanding arises that the "seen flash", which was initially recorded in our consciousness as a "sense experience", is now also understood as a (non-subjective) "event point" (Einstein, 2009). However, the idea (embodiment) of each point event is similar to what people think as soon as they refer to an "external reality world" (Einstein, 2009).

Albert Einstein (2009) and his readers learned that they are attracted to assigning a systematization of time to all experiences according to the method: If β is equal to α and Ω is equal to β, then Ω is also equal to α (succession of "experiences"). According to this way of thinking, what happens to all the point events that people are used to attributing to their experiences (Einstein, 2009)? The easiest way to think about this seems to be that the time systematization of events is similar to the time systematization of experiences (Einstein, 2009). Humans also did not do this specifically and without awareness until critical thought became valid (Einstein, 2009). For example, it is possible that an acoustic time systematization of experiences diverges with respect to a visual time systematization of events, which generally leads to the time systematization of events being difficult to discern from the time systematization of experiences (Einstein, 2009).

In order for a non-subjectivization of the (external world of our) reality to emerge, a further creative idea is required (Einstein, 2009): The event point also presents itself placed in a space, by no means solely within the time coordinate.

In the previous section, Albert Einstein (2009) and his readers attempted to understand how all constructs of time, space and event point can be inserted into a mental relationship with all experiences. Thinking illustrates these uninfluenced creations by the intelligence of man, methods for thinking that are suitable for organizing all experiences in a relationship and thereby keeping them in view more easily (Einstein, 2009). The experiment of learning these principles through empirical references will illustrate the extent to which humans truly exist dependent on these propositions (Einstein, 2009). Accordingly, people learn about their independence (in thinking), which is always a difficult process when there is a requirement to construct a suitable thought (Einstein, 2009).

With regard to the image of the mental origin of the constructs space-time-event point (Albert Einstein and his readers would like to refer to these in a shorter form as "space-related" in contrast to the constructs according to the mental world), Albert Einstein (2009) and his readers must also add descriptive things. Albert Einstein (2009) and his readers learned the spatial construct in conjunction with experiences with boxes and systematization of objective bodies within these boxes. The described comprehension design therefore already requires the construct (model conception) of an objective body (e.g. "box") (Einstein, 2009). In the same way, all people who were to be imagined for the design of a non-subjective understanding of time have also assumed their task as objective bodies within this system of relationships (Einstein, 2009). Albert Einstein (2009) therefore believed that before his readers could think of the model constructs of space and time, the design of the constructs of objective bodies had to take place.

The spatial constructs described above already belong to a popular way of thinking with regard to related constructs in the intellectual world, for example joy, start, reason, etc. (Einstein, 2009). With regard to the physical and overall scientific way of thinking, this way of thinking now seems to be characterized by the fact that it basically tries to use (think) all "space-related" constructs individually and aims to formulate any legal relation with these constructs (Einstein, 2009). The quantum physicists think sounds as well as colorations in terms of movements (frequencies) to reduce, the psychologists look for mind as well as concern in conspicuous processes, in such a way that the mental can be removed altogether from the causality matrix (causality nexus) of the existing (thinking), thus nowhere occurs as an independent node within the causality links (thought links) (Einstein, 2009). This understanding, which considers this summation of causality thoughts (connecting nodes) to be achievable with the exclusive use of merely "space-related" constructs according to this principle, is certainly similar to the reason why relationships are generally thought of today as "materiality" (relation between matter and thought) [previously, "substances" had this task as a basic construct (freedom between matter and thought)] (Einstein, 2009).

Why does it seem necessary to derive these principles of a natural science way of thinking from the supernatural realms according to Plato and to strive to recognize the same earthbound origin (Einstein, 2009)? Enlightenment (Einstein, 2009): So that the constructs are released regarding their attached prohibition, as well as thereby establish higher independence within the thought design (the sentence creation or construct design). Hume and Mach introduced this reflective criticism, which is like a lasting success attributable to both of them (Einstein, 2009).

Scientists have transferred, concretized and modified the constructs of objective bodies (including an important special idea of "solid bodies"), space and time from the general (popular) way of thinking (Einstein, 2009). Their first important result was the derivation of geometry according to Euclid (Einstein, 2009). Albert Einstein (2009) and his readers must pay attention to the associated initial (axiomatic) mode of expression, as this can conceal its empirical origin (placement locations of solids). According to the special sense, space with its three dimensions and its description according to Euclid (space can be filled with "cubes" of the same size without exception) also have an empirical origin (Einstein, 2009).

The fineness (subtlety) of the understanding of space increased with the realization that bodies never exist completely motionless (Einstein, 2009). Every body is flexibly deformable and deforms its filling volume when the temperature changes (Einstein, 2009). The shapes of those conceivable places that must appear to be selected by means of Euclid's geometry for shaping (describing) can therefore never be determined without the theories (equations) from physics (Einstein, 2009). However, because our physics already has to use geometry during its (mental) definition of the constructs, an empirical degree of this geometry can only be determined and evaluated within the limits of physics as a whole (Einstein, 2009).

According to the same allegory (empiricism to geometry AND geometry to physics), one should also think of atomism and its understanding of the non-infinite possibility of division (Einstein, 2009). Because subatomically extended spaces cannot be measured under any circumstances (Einstein, 2009). Likewise, atomism requires that the idea of clearly and non-dynamically defined wall surfaces of solids be dissolved according to this principle (Einstein, 2009). Accordingly, precisely understood, there are no independent theories about the placement locations of solids, not even in the empirical field of macroquantum physics (Einstein, 2009).

Nevertheless, no one concluded from this that the space construct should be dissolved, because this was necessary in a first-class scientific system of natural physics (Einstein, 2009). In the 19th century, it was Mach alone who reasonably intended to discard (or eliminate) the space construct by exchanging space by means of the construct of a summation of all present distances of each point of matter (Einstein, 2009). (Mach undertook the described train of thought because he wanted to gain an understanding of body inertia that was satisfactory in the opinion of Albert Einstein (2009)).

A spacetime field (Einstein, 2009). Within the classical theory, time and space have two tasks (Einstein, 2009). The first task is to be the influencer (carrier) or grid (frame) with respect to the physical events, with respect to which all point events are characterized by means of the spatial coordinates and the time coordinate (Einstein, 2009). All matter is thought to be composed of "matter points", the mechanics of which determine the physical events (Einstein, 2009). If the points of matter are imagined to be constant (continuous), then this occurs, so to speak, temporarily (provisionally) within these scenarios, within which scientists do not want to or cannot represent their discontinuous design (Einstein, 2009). Within this scenario, the smallest particles (volume components) of matter can be considered almost the same with regard to the points of matter, at least as long as this view only includes mechanics and no processes that previously seemed impracticable or nonsensical to explain mentally on the basis of mechanics (e.g. temperature shifts, chemical processes) (Einstein, 2009). Space and time had the second task of functioning mentally like an "inertial system" (Einstein, 2009). Inertial systems (body inertial systems) were given preference with regard to all conceivable coordinate systems (systems of reference bodies) because the principle of inertia demanded validity with regard to these reference systems (Einstein, 2009).

The meaning of this is similar to "physical reality", which is not dependent on those living beings (subjects) who think and experience this reality, which was to be understood as consisting of time and space on the one hand and, with regard to these, of moving, constantly existing points of matter on the other – at least according to the concept (Einstein, 2009). The idea of the free existence of time and space can generally be expressed figuratively (Einstein, 2009): If all matter were to disappear, then time as well as space would accordingly continue to exist individually (like an idea of a place within which physical events take place).

The improvement of the described understanding arose due to the establishment, which at first did not seem to be related to the question of the existence of (matter-)time-space – an emergence of a field construct as well as the same mental consequence of exchanging the particle idea (matter point) according to the concept (Einstein, 2009). Within the grid (thought matrix) of Newton's laws of physics, the field construct was accepted as a supporting idea, within scenarios where physicists considered all matter as one context (continuum) (Einstein, 2009). For example, during the observation of heat transfer within a solid, its status is described by the fact that within each point of the body, with respect to each determinable time coordinate, a temperature experience is related (Einstein, 2009). This means mathematically (Einstein, 2009): A temperature experience T describes analogous to a mathematical equation (property) of the space-related expansion (spatial coordination function) within the time coordinate t (field temperature). The physics of heat transfer resembles a local relation reference system (differential equation system) that expresses every special case of heat transfer (Einstein, 2009). The same temperature experience T resembles an easy-to-understand example of experience in terms of field understanding (or field concept) (Einstein, 2009). A field resembles a value [or a bundle of values (vector)] that represents the space-time coordination movement as a property (Einstein, 2009). Another exemplary experience is similar to the representation of fluid mechanics (Einstein, 2009). Within all fluid points (fluid particles), the specific velocity movement v exists at all times, which is quantitatively conceivable on the basis of its three "components" with regard to the coordinate system axes (four-vector) (Einstein, 2009). All velocity components in a (four-dimensional) point event (field components) also have the same properties in terms of spatial coordination (x, y, z) and time coordination (t) according to this experience of motion (Einstein, 2009).

With regard to the field concepts described (temperature field, fluid field), it appears to be decisive for them that these only arise within heavy (weighty, weighable, ponderable) point masses; the physicists only want to specify a status of matter (Einstein, 2009). In places where matter did not exist, no field continuum was assumed either – in agreement with the establishment anecdote of field understanding (Einstein, 2009). Now, however, physicists found in a first quartile related to the calendar years 1800 to 1900 that all superposition and curvature observations of the light rays could be justified with considerable accuracy if the light rays were generally understood to be analogous to a wave particle field that was completely similar to a moving vibration field within a flexible solid body (Einstein, 2009). Accordingly, physicists thought it was therefore necessary, due to the equation, to establish the field construct, which can also exist independently of heavy matter in a disembodied space (Einstein, 2009).

This fact created a situational paradox, as the understanding of fields, consistent with its origin, subsequently seemed limited to representing inertial states within a heavy solid (Einstein, 2009). This seemed the more correct to assume from the time that physicists shared the belief that all fields were analogous to moving relatable states, which required the presence of matter (Einstein, 2009). Accordingly, physicists thought it was necessary to also endorse the presence of matter at every location within space, which they called the "aether construct" (matter construct) (Einstein, 2009).

This independence of this field understanding from a theory about the necessity of supporting matter resembles one of the most decisive psychological processes within the evolution of physics thinking (Einstein, 2009). Within the third to fourth quartile with regard to the calendar years 1800 to 1900, after the publication of Faraday's and Maxwell's research results, there was an increasingly understandable view that a field-related explanation of electromagnetic processes appeared very progressive compared to the consideration based on point-moving ideas (Einstein, 2009). With the establishment of a field understanding within electrodynamics, Maxwell was able to predict the existence of electromagnetic particle waves, whose fundamental equality with respect to all light body waves already left no doubt due to an analogy with the speed of motion (Einstein, 2009). Consequently, electrodynamics integrated (absorbed) optics in its principle (Einstein, 2009). The intellectual effect of this great achievement was that the understanding of the field in terms of a physical grid of Newtonian mechanics gradually achieved greater freedom (Einstein, 2009).

Nevertheless, it was temporarily readily thought that all electromagnetic fields were analogous to the momentary states of the aether construct, and physicists enthusiastically thought of their momentary states as moving moments (Einstein, 2009). After their attempts at explanation always failed, physicists gradually accepted to think away this moving interpretation (as a false sense or interpretation) (Einstein, 2009). However, physicists still retained the view that all electromagnetic fields corresponded to momentary states of an aether construct; this was the state of affairs near the end of the next century (Einstein, 2009).

The assumed aether idea gave rise to the following question (Einstein, 2009): What is the behavior of the aether in its motion relation with respect to heavy bodies? Does the aether construct (itself) have the same body mechanics or are the aether components comparatively motionless (Einstein, 2009)? A large number of sophisticated experiments have been carried out to answer the latter question (Einstein, 2009). Within this context, important findings also appeared to become relevant, namely the aberration of all fixed stars as a consequence of the annual motion of our Earth and the "Doppler effect" (effect of the relative mechanics of all fixed stars with regard to the oscillation of the light rays reaching the Earth with a certain radiation oscillation) (Einstein, 2009). Lorentz substantiated all the results of the described facts and experiments (except for the Michelson-Morley experiment) according to the theory that the aether construct does not participate in the mechanics (motion) of heavy (weighty) bodies and that all components of the aether construct do not have relative mechanics (Einstein, 2009). The aether construct was, so to speak, thought to be equivalent to the nature of a completely motionless space (Einstein, 2009). However, this study by Lorentz still had an impact (Einstein, 2009). This (Lorentz's analysis) described all accepted electromagnetism processes and optics processes within heavy mass bodies at that time according to the theory that the effect of a heavy mass with respect to the electric field (as well as the opposite) appears to be due solely to the fact that all quanta of a mass comprise electric components that are involved within quantum mechanics (particle motion) (Einstein, 2009). Lorentz illustrated with regard to the experiment of the scientists Michelson and Morley that his result at least does not mean a contradiction with regard to an assumption about the immobile aether construct (Einstein, 2009).

Nevertheless, the development of the theory (of everything) was not yet completely satisfactory solely on the basis of all the appropriate thoughts, and precisely on the basis of the next reasoning (Einstein, 2009). Newton's motion, which should be confirmable with all certainty by its validity possible with high approximation, describes the doctrine of the equality of each inertial system (or inertial space) with the expression of each law of nature (scalar of natural laws regarding the transformation from one to the next inertial system) (Einstein, 2009). All electromagnetism and optics experiments made the same thing learnable with high precision (Einstein, 2009). However, the basis for the assumption about electromagnetism arose from the preference for a particular inertial reference frame, in a sense that of an unmoving aether light (Einstein, 2009). This understanding of this assumption-based foundation was still completely inadequate (Einstein, 2009). Was there no adaptation of this basis, which – analogous to Newtonian motion – can maintain an equality of inertial reference frames (special principle of relativity) (Einstein, 2009)?

A modification of Newton's theory, as an answer to this question, is similar to the special theory of relativity (Einstein, 2009). The theory integrates a condition from the Maxwell-Lorentz assumption for the uniformity of the speed of light bodies within an empty space (Einstein, 2009). In order for the (rectilinear) uniformity of motion (of light) to coincide with the uniformity of the inertial reference frames (special relativity principle), there can be no unrestricted expression of simultaneity; in addition, transitions according to Lorentz with respect to the time coordinate and spatial coordination with respect to the transformation from one inertial reference frame to the other (inertial reference frame) (Einstein, 2009). An entire sense of the special theory of relativity appears integrated within the principle (Einstein, 2009): All laws of nature behave constantly (like the speed of light) with respect to all Lorentz transitions. The meaning of this expression is that the theory establishes (or limits) all conceivable laws of nature with a coincidentally selected (specific) method (Einstein, 2009).

In what way does Albert Einstein (2009) understand (comprehend) the state of space in his special theory of relativity? In the first place, it is generally accepted that all four dimensions of our reality were already established before this theory of relativity (Einstein, 2009). Also within Newton's motion, an event point with four values is found locally, so to speak with three spatial coordinates and one time coordinate; a summation of all "event points" is therefore physically enclosed within a multiplicity that remains constant in four dimensions (Einstein, 2009). However, according to Newton's theory of motion, the space continuum consisting of four dimensions is divided objectively into a time on one dimension and into spatial forms consisting of three dimensions, which, with regard to the continuum, only include event points considered with simultaneity (Einstein, 2009). This division applies equally to all inertial systems (Einstein, 2009). A simultaneity existing between two selected event points considered as an inertial frame of reference occurs (mentally) simultaneously with respect to the simultaneity of these event points (or these inertial frames of reference) with respect to all (also simultaneously existing) inertial frames of reference (Einstein, 2009). This is similar to the opinion, if generally expressed, that time exists universally (i.e. absolutely in the space continuum) in Newtonian motion (Einstein, 2009). According to Albert Einstein's (2009) special theory of relativity, time exists differently. The basic idea of event points, which exist in simultaneity with the event (system) under consideration, naturally applies with regard to a selected inertial system, but now depends on a selection (determinacy) of an inertial system (Einstein, 2009). A worldview with four dimensions now remains a single objective body that includes each event point existing in simultaneity separately; a "today" (or time) no longer has any valid objectivity for a geometrically expanded space continuum (Einstein, 2009). This is similar to the relationship that, in general, a time coordinate and the (three) space coordinates are to be understood inseparably as a world with four dimensions because of their objectivity, if in general the sense of objective relations must be formulated detached from a negligible classical arbitrariness (Einstein, 2009).

Because Albert Einstein's (2009) special theory of relativity physically explained an equality of inertial systems, it evaluated the contradiction in the therefore untenable assumption regarding an unmoving aether construct. Physicists were therefore prompted to abandon the idea that an electromagnetic field could be understood as having the same status as a supporting matter (Einstein, 2009). A field construct thus appears as an indecomposable component of a physics-based explanation, indecomposable with the same meaning with respect to the matter construct within the classical theory of motion (Einstein, 2009).

So far, Albert Einstein (2009) and his readers have focused their attention on the ways in which our understandings of time and space have been altered by special relativity. Now, however, Albert Einstein (2009) and his readers would like to consider the components that special relativity has removed from Newtonian motion. Also in this theory, all laws of nature are only valid if a space-time representation has an inertial frame as its basis (Einstein, 2009). An inertial principle and the principle regarding the uniformity of the speed of light bodies must only be valid with regard to an inertial system (Einstein, 2009). Likewise, all field theories require significance and validity only with regard to the inertial systems (Einstein, 2009). Analogous to Newtonian motion, a spatial construct in the special theory of relativity is therefore also an independent component of a description of physics-based reality (Einstein, 2009). An (inertial) space (system) – or more precisely, the (objectified) space (body) linked with respect to a localized time coordinate – continues to exist if the field construct and the matter construct are generally considered to be distant (Einstein, 2009). This physical construct with four dimensions (Minkowski's space) is conceptually similar to a support for all matter and a field (Einstein, 2009). The inertial space systems, including the moments of time localized there, merely resemble privileged reference systems with four coordinate dimensions, which exist connected to each other by means of Lorentz's rectilinear transitions (transformations) (Einstein, 2009). Because there are no other components within this physical construct with four dimensions that represent an objectivity of the "today", the understanding of the event and its emergence does not, of course, experience a complete abolition, but nevertheless a complexity (Einstein, 2009). Therefore, it seems more consistent with nature to equate physical reality with an existence on four dimensions (x, y, z, t) instead of the emergence of this existence on three dimensions (x, y, z) (Einstein, 2009).

The (almost) stationary (practically rigid) space with four dimensions described in the special theory of relativity is similar to the four dimensions of the stationary (rigid) aether construct with three dimensions according to Lorentz (Einstein, 2009). The expression is also valid with regard to the special theory of relativity (Einstein, 2009): Space has been assumed to be present as well as independently existing for the physical description of states since the Big Bang. Accordingly, the special theory of relativity in no way resolves Descartes' concern regarding an independent, truly non-A-posteriori existence of a "contentless space" (Einstein, 2009). The way in which this limitation is to be triggered by the general theory of relativity corresponds to the actual objective of the fundamental ideas listed in this section (Einstein, 2009).

A spatial state within Albert Einstein's general theory of relativity (Einstein, 2009). Albert Einstein's (2009) theory of relativity arises in particular from the derivation of his goal (motivation) of being able to understand a correspondence between mass inertia and mass gravity. In general, an inertial system K_1 is assumed, which according to physics resembles an empty space (Einstein, 2009). This means that within the considered spatial component (K_1) there is neither a field concept according to Albert Einstein's (2009) special theory of relativity nor a concept of matter (according to the classical meaning). With regard to K_1, the second coordinate system K_2 is said to be uniformly moved (Einstein, 2009). K_2 therefore does not resemble an inertial system (Einstein, 2009). With respect to K_2 accelerated, all test masses would be moved and, of course, unaffected with respect to their state according to physics and chemistry (Einstein, 2009). With regard to K_2, there is therefore a status that generally – at least to a first approximation – does not allow any difference with regard to a gravitational field (Einstein, 2009). This observable fact is therefore consistent with the consideration (Einstein, 2009): K_2 also appears to be an "inertial system" in the same sense, but with respect to K_2 the (uniform) gravitational field exists (within this relation the origin of the gravitational field is mentally neglected). As soon as a gravitational field is generally included within the grid (frame matrix) of this view, then an inertial system no longer has any objective meaning, but only if the described "principle of equality" (equivalence concept) applies as a consequence extended for an arbitrary relative mechanics of the coordinate systems (Einstein, 2009). If it seems conceivable to construct a theory that is free of contradictions with regard to this principle, then this theory (automatically) resembles an experience-based truth about the equivalence of mass inertia and mass gravity on the basis of the conceptual design (Einstein, 2009).

Observed with four coordinates, the transformation between K_1 and K_2 resembles a curvilinear transition based on four dimensions (Einstein, 2009). This now leads to a question (Einstein, 2009): What kind of curvilinear transitions can be generally accepted or in what way can a transformation be generalized according to Lorentz? In order to find an answer to this question, the next thought appears to be decisive (Einstein, 2009).

An inertial system corresponding to the preceding theories of relativity (classical and special mechanics) is assigned a function (Einstein, 2009): Dimensional differences (coordinate distances) are measurable with (unmoving) "rigid" test rods. Further dimensional differences (time intervals) with (unmoving) "rigid" clock hands (Einstein, 2009). This idea of properties is extended by the idea that with regard to multiplicity relative to the positioning of immobile unit rods, all axioms regarding these "distances" are valid according to Euclid's geometry (Einstein, 2009). With the help of the results of Albert Einstein's (2009) special theory of relativity, a consequence generally arises subsequently due to a fundamental way of thinking that such a direct thought assumption disappears according to physics with regard to coordinate dimensions of line element systems (K_2) moving relative to inertial systems (K_1). However, if this physics appears true, [that no transitions (transformations) exist,] then all point coordinates are merely more like a systematization of "being adjacent" (or coexistence) [and thus also a dimensional extent (curvature intensity) of our space], but not measurable functions of our space (Einstein, 2009). In general, accordingly, it becomes conceivable to continue all transitions in terms of arbitrary infinite dimensionings; transformations as a concept is not an exact equation form here (Einstein, 2009). This involves the general principle of relativity (Einstein, 2009). All laws of nature can only exist covariantly with regard to arbitrary continuous coordinate transformations (for example, space-time K_1 can be neighboring but nevertheless differentially accelerated to space-time K_2) (Einstein, 2009). The described consequence (in the context of a sequence of conceivable logical clarity of laws of nature) limits all possible (general) laws of nature according to this condition to different degrees, whereby a coincident equation with the special principle of relativity becomes unthinkable (Einstein, 2009).

This thought process is largely confirmed by an understanding of the field analogous to an independent construct (Einstein, 2009). Since all laws of nature valid with respect to K_2 are (generally) to be understood analogously to a gravitational field, a question regarding the existence of matter, which causes the field property, is neglected (Einstein, 2009). This thought process also makes the laws comprehensible as to why all laws of nature concerning the matter-free gravitational field are directly connected with the incidence via the general spatial relativity in the same way as the laws of nature concerning fields of a generally valid class (if, for example, the electromagnetic field exists) (Einstein, 2009). Albert Einstein (2009) and his readers have, coincident with experience, a suitable justification for their theory, so that a "fieldless" Minkowski space concept represents a special case conceivable according to the laws of nature, and of course a special case that is as simple as possible to understand. This concept of space is described in terms of the attributed measurable function (metric), so that $dx_1^2 + dx_2^2 + dx_3^2$ resembles a square consisting of the space-based distance between two small (infinitesimal) adjacent event points on a spatial transverse surface in three dimensions (triangle) (Pythagoras' theorem), which is plotted with the unit metric (scale), while dx_4^2 resembles a time-based distance between two point events with shared spatial coordinates (x_1, x_2, x_3), which is plotted with a coincident time metric (Einstein, 2009). This all leads – which can be easily understood on the basis of transitions (transformations) according to Lorentz – to the fact that this value ds^2 (30) represents a meaningful metric due to its objectivity (Einstein, 2009).

(30) $ds^2 = dx_1^2 + dx_2^2 + dx_3^2 - dx_4^2$ (Einstein, 2009)

According to the mathematics of the constellation (i.e. the position of the stars in relation to each other), this metric is similar to the fact that the value ds^2 (30) represents an invariant or a scalar with regard to the Lorentz transitions as a scale; [this means that only the distances between the stars as point neighbors can be much larger or smaller] (Einstein, 2009).

If, in accordance with the general principle of relativity, an arbitrary continuous (infinitely small) transition of the point coordinates is now generally assumed to exist for the spatial state, then the value that is meaningful due to its objectivity can be expressed in a renewed reference system (with changed coordinate distances) with the aid of the relation (31), for which, with regard to all specifications (indices) i and k, each connection (or each transition or transformation) of the model tensor 11, 12, 13, 14, 21, 22, 23, 24, 31, 32, 33, 34, 41, 42, 43 and 44 must be added up (four-vector) (Einstein, 2009).

(31) $ds^2 = g_{ik}dx_idx_k$ (Einstein, 2009)

However, all g_{ik} now do not resemble any invariant, but instead properties of all point coordinates that are selected using the arbitrarily determined coordinate transformation (Einstein, 2009). Nevertheless, all g_{ik} do not resemble any properties of these modified point coordinates, but instead of course these properties, so that the system of equations (31) is changeable (transformable) with the help of a continuous coordinate transformation to four dimensions back to the system of equations (30) (Einstein, 2009). In order for this to be conceivable, all properties must comply with certain generally "covariant" precondition equation systems, which Riemann established more than 50 years ago, even before the derivation of Albert Einstein's (2009) general theory of relativity ("Riemann condition"). In accordance with the principle of equality (equivalence axiom), (31) models the gravitational field within the special case within a general covariant expression if all g_{ik} comply with this "Riemann condition" (Einstein, 2009).

The theory with respect to an independent gravitational field as a general case must therefore fulfill the next condition (Einstein, 2009). The theory of gravitation exists in accordance with experience if this "Riemann condition" is met; however, this theory has a smaller scope (mentally compared to π) and is therefore less restrictive than a "Riemann condition" (Einstein, 2009). Thus, a theory of an independent gravitational field is almost completely defined, the exact derivation of which does not require a deeper explanation in this section (Einstein, 2009).

Now Albert Einstein (2009) and his readers have been able to understand how the transformation (from special relativity to general relativity) changes a state of space. Consistent with Newtonian theory and special relativity mechanics, a space (space-inseparable-from-time, space-time for short) has an independent being compared to the concept of matter or field (Einstein, 2009). In order for this space-filling(-something), which is independent with regard to the point coordinates, to be able to be experienced, the space-time system or inertial system must have been understood as existing since the Big Bang with the help of its characteristic metrics, since in the opposite case no model representation of this "space-filling" would be conceivable (Einstein, 2009). If the general way of thinking is followed that there is no space-filling (e.g. no space-field) mentally, then a measurable spatial state according to (30) always exists afterwards as before, which would also express the inertial behavior of a test reference body moved into this space (Einstein, 2009). Coincident with the general theory of relativity, a space has no special being with respect to a "space-filling", with respect to the point coordinates non-independent (Einstein, 2009). In general, for example, the independent gravitational field can be expressed by means of all g_{ik} (analogous to coordinate properties) by transforming all systems of equations modeling the gravitational field (Einstein, 2009). As soon as a gravitational field, i.e. all properties g_{ik}, is absent in general, then the spatial state of kind (30) no longer exists, instead there is no space at all, nor is there a "spatial topology" (i.e. no spaces within spaces etc.) (Einstein, 2009). In accordance with experience, all properties of g_{ik} do not merely determine a field, but also, in simultaneity, all metric and topological design possibilities of (spatial) multiplicity (Einstein, 2009). In the general theory of relativity, the spatial state of type (30) does not resemble a spatial state with a missing field state, so to speak, but rather a special case of a g_{ik} field, with respect to which all g_{ik} [with respect to the selected reference system (mentally systematized idea of coordinates), to which no objective sense can be assigned] have quantities that are independent with respect to the point coordinates; a spatial state without content, i.e. a spatial state with a missing field state, cannot exist under any circumstances (Einstein, 2009).

So Descartes was right (with his truth) when he said that it must be necessary to give up the existence of contentless space (Einstein, 2009). Of course, this view seems incomprehensible until the reality that can be perceived according to physics is no longer only mentally visible within heavy bodies (Einstein, 2009). The idea of a field analogous to a carrier of reality, linked with the help of the general principle of relativity, illustrates the actual truth of Descartes' thought for the first time (Einstein, 2009): A "field-less" spatial state does not exist.

General gravitational field theory (Einstein, 2009). An assumption about the independent gravitational field on the basis of Albert Einstein's general theory of relativity seems easy to understand because Albert Einstein (2009) and his readers are allowed to (mentally) connect to it that a "fieldless" Minkowski space state has to coincide with a metric property according to (30) of the generalized gravitational field theory. With the help of this special case, gravitational theory arises as a consequence of a generalization to which almost no arbitrariness is attached (Einstein, 2009). A subsequent extension of this theory of gravitation does not appear to be predetermined by the general principle of relativity; the further development of this theory was tried out in the previous century with different approaches (Einstein, 2009). Each of these thought experiments shares the following view: physical reality is to be understood analogously to a field, this field represents the generalization of our gravitational field, the field theory represents the generalization of the theory of an independent gravitational field (Einstein, 2009). Albert Einstein (2009) now assumed that after many attempts he could write down a most real way of expression with regard to generalization in an understandable way, but he considered himself incapable of thinking up whether this general gravitational field theory could exist confronted with all possible experiences.

A generalization can generally be described as follows (with respect to a gravitational field, i.e. the field of gravity) (Einstein, 2009). An independent gravitational field of all g_{ik} has a symmetric function g_{ik} equal to g_{ki} (g_{34} equal to g_{43} etc.) in accordance with the underlying induction (derivation) using the contentless "Minkowski space" (Einstein, 2009). A field in the general case appears with the same properties, but with an anti-symmetric function g_{ik} not equal to g_{ki} (g_{34} not equal to g_{43} etc.). A deduction (derivation) of gravitational field theory is completely similar to the special case of an independent gravitational field (Einstein, 2009).

With regard to the previous general consideration, a question concerning the special field theory appears to be of secondary importance (Einstein, 2009). This most important question today is whether the gravitational field laws can simply lead to the truth with regard to their nature considered in this section (Einstein, 2009). Albert Einstein (2009) refers here to gravitational field theory, which can fully represent physical reality (taking into account a spatial state with four dimensions) with the help of a field. Our current generation of physicists tends not to be able to answer this (most important) question with a yes; this generation of physicists believes, as a consequence of the way this quantum mechanics is expressed today, that a state of a space (system) can by no means be described directly (as Albert Einstein believes), but instead only directly by means of the representation of numerical statistics of all measurement results that can be experienced within this space (system); the prevailing opinion has the effect of (mentally) limiting (interpreting) that a dualistic light nature (body wave design) evaluated experimentally could only be experienced with a corresponding minimum level of understanding of reality (Einstein, 2009). Albert Einstein (2009) theorizes that such an extensive conceivable rejection of knowledge (logic problem) by means of our real experience (which is similar to today's knowledge) now exists (in the sense) without logic (or as a falsity) and that the general public (physicists or humanity in general) must think of this as a challenge to further develop the method of gravitational field relativity theory to completion (completion).

Second section: On the general relativity theory

6 Special and general relativity principle

The physical principle of special relativity applies to all uniform motion (Einstein, 2009). Every movement represents a relative movement corresponding to the general understanding (Einstein, 2009). According to Albert Einstein (2009), the motion of two bodies can be expressed analogously in two uniform relative motions: (1) the body K is moving relative to the body K' and (2) the body K' is moving relative to the body K, i.e. both bodies are moving in inverse relativity to each other.

For statement (1), the body K' applies, for statement (2), the body K applies as the reference system (Einstein, 2009). The selection of a reference system with regard to the motion of a body generally has no influence on the sole representation of the relative motion (Einstein, 2009). The relative motion of a number of bodies in relation to each other is given by nature in the narrower sense, but this must not be confused with the principle of relativity in the broader sense, which served as a basis for the previous analyses (Sections 21 to 37) (Einstein, 2009).

The principle outlined above implies that the body K as well as the body K' could be chosen as the reference system for the representation of any event (since this corresponds to nature) (Einstein, 2009). According to Albert Einstein (2009), this principle also includes the following: If general laws of nature are formulated on the basis of experience by choosing (1) the body K' as the reference system or (2) the body K as the reference system, then laws with exactly the same meaning are generally determined for both expressions, for example for mechanics or the speed of light in a vacuum. In simpler terms (Einstein, 2009): No reference system can be preferred to a different reference system, such as K to K' or K' to K, for the physical representation of generally existing natural processes. The correctness or incorrectness of this principle in the broader sense can only be decided by a researcher through experience, but this does not mean that its content must necessarily be correct or incorrect in advance like the principle in the narrower sense, since the properties direction of motion and reference system do not occur in the latter (Einstein, 2009).

So far, the indifference of each reference system K with respect to the representation of general laws of nature has not been expressed (Einstein, 2009). The approach was more like the following (Einstein, 2009). Albert Einstein (2009) first hypothesized that a reference system K with a certain state of motion exists, so that relative to it the Galilean principle is valid: a single independent point mass sufficiently distant from any other mass propagates in a straight line and uniformly. The general laws of nature should simply exist with respect to the Galilean reference system K (Einstein, 2009). In addition to K, any reference system K' would have to be preferable and, like K, exactly equivalent for the representation of general laws of nature that have a uniform and rectilinear motion relative to K without rotation: Each of these reference systems is to be called a Galilean reference system (Einstein, 2009). The principle of relativity is only valid for Galilean reference frames, but not for reference frames that deviate from it due to other motion (Einstein, 2009). In terms of content, reference is made here to the special principle of relativity or the special theory of relativity (Einstein, 2009).

On the contrary, the general principle of relativity is understood as the assumption: Every reference system, such as K and K' etc., can be equally invoked to represent nature (modeling and visualizing general laws of nature), regardless of its state of motion (Einstein, 2009). At this point, it must be noted immediately that, due to physical facts that are not yet important, the assumption just made is replaced with a more general assumption (Einstein, 2009).

Since the special principle of relativity was established after its publication, it should appeal to every mind to research an abstraction (generalization), i.e. to attempt a further development towards a more general principle of relativity (Einstein, 2009). Such an attempt initially seems impossible due to a comprehensible, probably entirely accurate investigation (Einstein, 2009). A reader should imagine himself inside a uniformly moving body K (Einstein, 2009). If K is always moving uniformly, the reader will not notice the motion of the body (Einstein, 2009). It is therefore also the case that the reader can understand the physical fact without inner aversion to the effect that the body is at rest, but that another body K' must be in motion (Einstein, 2009). This understanding is of course also completely correct from a physical point of view in accordance with the special principle of relativity (Einstein, 2009).

If the uniform motion of the body K is now changed into a non-uniform motion by violently decelerating the body K, then an internal reader in the same direction of forward motion experiences an equally violent thrust (Einstein, 2009). An accelerating movement of a body K shows itself relative to this in the body behavior according to mechanics, because the behavior according to mechanics is different from the example illustrated above, therefore it seems as if it is impossible that the same laws of mechanics are valid relative to the non-uniformly moving body K and relative to the uniformly moving, i.e. resting body K (Einstein, 2009). Now it is understandable that the Galilean principle is invalid relative to non-uniformly moving reference systems (Einstein, 2009). Therefore, it is first necessary to assign a class of physical non-dynamic reality to non-uniformly moving reference bodies, contrary to the general principle of relativity (Einstein, 2009). However, it will soon be established that this conclusion is not convincing (Einstein, 2009).

7 The field of gravity

A question such as: "Why do bodies that are lifted up and then released fall back down to a planet like the Earth?" usually involves the answer of classical physics: "Because they experience an attraction from the mass of a planet like the Earth" (Einstein, 2009). This answer is expressed slightly differently by more modern physics due to the next argument (Einstein, 2009). Through detailed study of electromagnetic phenomena, physicists have formed the opinion that such an interaction does not exist directly in relation to distance (Einstein, 2009). For example, if a piece of iron is attracted by a magnet, then readers should not accept the opinion that a magnet directly influences the piece of iron through the free space in between, instead, according to Faraday, readers should imagine that something real is physically caused continuously in the space around this magnet, which is called a magnetic field (Einstein, 2009). This field itself has a magnetic effect on the piece of iron, whereby the iron attempts to propagate with respect to the magnet (Einstein, 2009). Whether the magnetic field represents an intermediate definition (for the field of gravitation) will not be discussed here, as its claim (to be a definition) is actually interchangeable (Einstein, 2009). It should only be noted that readers can use this intermediate definition to model and visualize electromagnetic phenomena, especially the movement of electromagnetic waves, much better on the basis of theory than without an intermediate definition (Einstein, 2009). Readers can also understand the interaction of gravity in the same way (Einstein, 2009).

The influence of a planet, such as the Earth, on a body is not immediate (Einstein, 2009). A planet produces a gravitational field in its spatial proximity (Einstein, 2009). Its gravitational field influences a body and is the reason for its gravitational acceleration (Einstein, 2009). In accordance with experience, the strength of influence or interaction on a reference system decreases as soon as the distance from a planet, such as the Earth, is gradually increased, in accordance with a special law of nature (Einstein, 2009). (Albert Einstein (2009) makes a science-based prediction at this point, as space travel had not yet been tested in practice in 1916, the year in which the general theory of relativity was finalized). This law of nature can be formulated to fit the following assumption (Einstein, 2009): For the correct modelling and visualization of the reduction of the interaction of gravity with increasing distance from the acting reference body, the spatial properties of the gravitational field must be applicable, which should be able to depict a completely special law of nature. This suggests the idea that a reference body, such as the Earth, directly produces the gravitational field in its immediate spatial environment, its intensity and direction of action with increasing distance are therefore given by the law of nature, which contains the spatial properties of the gravitational field itself (Einstein, 2009).

In contrast to the gravitational field, the electric and magnetic field has no surprisingly unexpected property that is of fundamental importance for a law of nature (Einstein, 2009). The acceleration of bodies is not weakly related to their matter and physical properties, but solely to the interaction of gravity, the gravitational field (Einstein, 2009). For example, according to Einstein (2009), a piece of iron and a piece of plastic fall in this gravitational field (in a vacuum) in exactly the same way, regardless of whether they are dropped with or without the same initial speed. This highly precisely recognized law of nature can also be described in a different way with the help of the subsequent observation (Einstein, 2009).

Newton's law of motion states that the

(interaction) = (mass inertia) × (body acceleration),

here, inertia is a characteristic invariant of accelerated bodies (Einstein, 2009). If, according to Albert Einstein (2009), the driving interaction corresponds to gravity, i.e. the gravitational field, then the

(interaction) = (mass gravity) × (gravity field strength),

the mass gravity is also a characteristic invariant for the accelerated bodies. With the help of these two relationships, it can be concluded according to Albert Einstein (2009):

(body acceleration) = (mass gravity)/(mass inertia) × (gravitational field strength).

If it is now assumed, in accordance with experience, that in an existing gravitational field the body acceleration is not dependent on the type and physical nature of bodies but is always the same, then the relationship between the mass inertia and mass gravity is also always the same for every body (Einstein, 2009). This relationship can be set equal to 1 by a suitable selection of quantities, so the law of nature is valid: The mass gravity and mass inertia of bodies are (always) equal to each other (Einstein, 2009).

However, classical (Newtonian) mechanics noticed this central law of nature, but made no statement about it (Einstein, 2009). If a satisfactory statement is to be made, it must be recognized (Einstein, 2009): An equal property of bodies describes their mass inertia or mass gravity depending on the situation. The extent to which this law of nature really applies and how its statement is linked to the general principle of relativity will be explained in the following section (Einstein, 2009).

8 The equation of mass inertia with mass gravity as a justification of the general relativity principle

If a large cube of unfilled space is conceived (by means of the coordinate values x, y, z, t), as far away as possible from large masses such as stars, so that the statement given by the Galilean principle (Einstein, 2009) applies with sufficient precision. This makes it conceivable to select a Galilean reference system for this part of the cosmos, relative to which immovable events are immovable and non-stationary points are always in uniform rectilinear motion (Einstein, 2009).

Another reference system can be selected as an extended yet smaller cube projected into the described cosmic cube in the form of a space in which an observer is present and has devices at his disposal (Einstein, 2009). For the observer, of course, the mass gravity does not exist (Einstein, 2009). In order not to float leisurely towards the ceiling with the slightest push with respect to the floor of the room, the observer has to tie himself to the floor of the room with ropes (Einstein, 2009).

A bracket with a rope is attached to the center of the room ceiling and a reader begins to pull on it with constant force (Einstein, 2009). The room, including the observer, then starts a uniformly accelerated upward movement (Einstein, 2009). If this upward movement is estimated starting from another reference system to which no rope is attached for pulling, the observer's traveling speed should increase over time to unbelievable levels (Einstein, 2009).

How, on the other hand, is this process evaluated by the observer in the room (Einstein, 2009)? The acceleration in space is transferred to the observer by the floor of the room by means of counteraction (Einstein, 2009). The observer should therefore compensate for this effect with his legs if he does not want to feel the floor of the room with the entire length of his body (Einstein, 2009). This observer therefore stands in space in the same way as another observer in a room on a planet, such as the Earth (Einstein, 2009). If the observer standing in space drops a body from his hand, the spatial acceleration acting on the body should stop, therefore the body will adjust itself in accelerated motion (downwards) relative to the floor of the room (Einstein, 2009). This observer should also notice in space that the relative acceleration of the body with respect to the floor of the room always remains constant, regardless of the body he or she wants to drop in order to perform this experiment (Einstein, 2009).

Accordingly, an observer in space, supported by his knowledge of the gravitational field as explained in the previous section, should realize that, including space, he or she is in an extremely constant gravitational field (Einstein, 2009). This observer should, however, be puzzled for a moment by the fact that space (itself) does not fall down in the gravitational field (Einstein, 2009). However, this observer then finds the support in the middle of the room and the taut rope attached to it, whereupon he or she realizes that the room is immobile, i.e. at rest, in the gravitational field (Einstein, 2009).

Is it permissible to laugh at this observer and say that he or she has a wrong opinion (Einstein, 2009)? It must be assumed that this is not allowed if the reasoning is to remain unchanged, but it must be admitted that his or her opinion does not represent a violation of reason and the existing laws of mechanics (Einstein, 2009). Space, even if it is accelerated with respect to the Galilean space first illustrated, can still be considered immobile (Einstein, 2009). There is therefore a comprehensible reason to extend the special principle of relativity to reference systems that are accelerated relative to each other and Albert Einstein (2009) and his readers therefore have a strong reason to develop a more general principle of relativity.

The potential of this theory extension to be considered is based on the fundamental effects (properties) of the gravitational field, whereby every body experiences an equal rate of acceleration, i.e. the law of equality of inertia and mass gravity (Einstein, 2009). Since this law exists in nature, an observer in an accelerated space can understand the fundamental behavior of bodies in his or her environment through the properties of the gravitational field, and he or she is able, based on experience, to assume their frame of reference as a stationary system (Einstein, 2009).

An observer inside the room attached a rope to the ceiling of the room and the other side of the rope to a body (Einstein, 2009). The tethered body causes the rope to hang vertically taut (Einstein, 2009). What is the cause of this rope tension (Einstein, 2009)? The observer standing within this space must assume: A body hanging on the rope experiences an interaction in the direction of the gravitational field, an equilibrium is maintained to this effect by the rope tension (tensile force) (Einstein, 2009). The mass of the attached body is decisive for the magnitude of this kinetic tension (Einstein, 2009). Another observer flying freely outside this space, on the other hand, must assume the same: The rope must withstand the accelerated spatial motion and transfers this acceleration to the body attached to it (Einstein, 2009). The tension of the rope is correspondingly high, so that this (tensile force) can directly cause the acceleration of the body (Einstein, 2009). The mass inertia of the attached body is decisive for the magnitude of this rope tension (Einstein, 2009). Based on the observation example, it can be seen that the law of equality between mass inertia and mass gravity is required by the theory extension to the principle of relativity (Einstein, 2009). This has led to a physical understanding of this law (Einstein, 2009).

From an observation of the acceleration of space, it is learned that the more general theory of relativity has to provide central insights into gravitational laws (Einstein, 2009). In fact, the continuous application of a more general relativity theorem yielded those gravitational laws that describe this gravitational field (Einstein, 2009). Already at this point, Albert Einstein (2009) warned the reader against a false understanding, which is to be explained by the following descriptions. For an observer in space, a gravitational field exists, even if the gravitational field does not exist for the reference system selected first (Einstein, 2009). Readers could now simply come to the conclusion that it always merely appears as if the existence of gravitational fields exists (Einstein, 2009). Readers might think that no matter what kind of gravitational field might exist, they could always choose a different reference frame accordingly, so that no gravitational field exists with respect to this body (Einstein, 2009). However, this is by no means true for every gravitational field, but only for gravitational fields with a completely special structure (Einstein, 2009). For example, it is then not possible to select a reference system so that the gravitational field of a planet, such as that of the Earth, (in the entire propagation) dissolves starting from this body (Einstein, 2009).

We now understand why the justification mentioned at the end of section 6 does not constitute an evaluation of the more general principle of relativity (Einstein, 2009). It is correct that an observer in the decelerated body experiences a thrust forward due to this deceleration, and that he or she notices the non-uniform body movement as a result (Einstein, 2009). Nevertheless, no one urges the observer to associate this thrust with an actual acceleration of the body (Einstein, 2009). The observer is also able to understand the experienced event as follows: The reference system in which I find myself is always immobile (Einstein, 2009). In reality, however, there is (at the same time as the deceleration period) a gravitational field with respect to my reference system that is not chronologically constant in the forward direction (Einstein, 2009). The effect of this gravitational field leads to a non-uniform movement of a body together with its planet, such as the Earth, in such a way that the previous velocity of the body in the backward direction continues to decrease (Einstein, 2009). This gravitational field is responsible for triggering the thrust of the observer (forwards) (Einstein, 2009).

9 To what extent are there principles of Newtonian mechanics and special relativity that require a general relativity theory?

To repeat, Newtonian mechanics is based on the law: If there is a sufficiently large distance between bodies of matter, bodies of matter propagate uniformly and in a straight line or remain at rest (Einstein, 2009). This principle can only be valid for reference systems K with certain assumed speeds of movement at which reference systems are in uniform motion (translation) relative to others (Einstein, 2009). This theorem is not valid for reference systems K' moving relative to each other (Einstein, 2009). Readers of Newtonian mechanics and special relativity can therefore distinguish between reference systems K, relative to which the laws of nature apply, and reference systems K', relative to which the laws of nature are invalid (Einstein, 2009).

However, no reasoning person should accept this principle (Einstein, 2009). Readers should ask questions like (Einstein, 2009): Why can it be that some reference frames K (or their velocities of motion) definitely exist before reference frames K' (or their velocities of motion)? What is the justification for this preference (Einstein, 2009)? To help readers understand how the last question is meant, a physical comparison is described (Einstein, 2009).

Albert Einstein (2009) stands in front of his stove with gas. There are two pots together on his stove that look almost identical to each other (Einstein, 2009). The two pots are half full of water (Einstein, 2009). Albert Einstein (2009) notices that there is no steam coming out of the first pot, but there is non-stop steam coming out of the second pot. Albert Einstein (2009) is surprised by this, just as he was if he had never seen a gas stove and a saucepan before. Albert Einstein (2009) now sees a shimmering blue thing under the second pot and nothing under the first pot, so his surprise is also reduced if he has never seen a gas flame before. Therefore, Albert Einstein (2009) can only assume that the blue thing will cause the escape of the water vapor, or at least perhaps cause it. However, if Albert Einstein (2009) does not see the blue thing under either of the two saucepans, and if he sees that the second saucepan steams continuously and the first does not, then he is surprised and dissatisfied as soon as he sees any fact that he can use as an explanation for the different behavior of these two saucepans.

In the same way, Albert Einstein (2009) searched in Newtonian mechanics (and in the special theory of relativity) without success for real things with which he would have been able to explain the different behaviors of bodies with regard to the coordinate systems K and K'. Newton already recognized the same limitation (restriction) and also tried to remove it without success (Einstein, 2009). However, E. Mach recognized this limitation most comprehensibly and therefore demanded that a new principle was needed for Newtonian mechanics (Einstein, 2009). This limitation can only be compensated for with the help of (modern) physics that follows the general principle of relativity (Einstein, 2009). In this (modern) physics, the formulas of the general theory of relativity are valid for all reference systems, regardless of their speed of movement (Einstein, 2009).

10 Several conclusions using the general relativity principle

For readers, the descriptions in section 8 showed that the general principle of relativity enables mankind to determine functions (properties) of the gravitational field using a completely theoretical approach (Einstein, 2009). For example, if the space-time sequence of any natural phenomenon is clear and how it behaves on a Galilean surface relative to a Galilean reference system K (Einstein, 2009). Whereupon readers can determine by means of completely theoretical procedures (calculations) how this clarified natural phenomenon behaves as seen from a reference system K' that is accelerated relative to K (Einstein, 2009). However, because a gravitational field (now) exists relative to this renewed reference frame K', readers then learn during the (mathematical) presentation how a gravitational field can influence the natural process under investigation (Einstein, 2009).

Then readers learn, for example, that bodies that perform uniform rectilinear movements relative to each other with respect to K (equal to Galileo's law) generally perform curvilinear accelerated movements relative to each other with respect to accelerated reference systems K' (space) (Einstein, 2009). The curved acceleration or accelerated curvature is similar to the effect of the gravitational field existing relative to K' with respect to propagated bodies (Einstein, 2009). It is clear that the gravitational field influences the movement of the body in this way, and this means that the observation is fundamentally not innovative (Einstein, 2009).

However, readers will gain an innovative insight of fundamental importance if they carry out the described observation for light (Einstein, 2009). Light moves in a straight line with its speed c relative to the Galilean reference system K (Einstein, 2009). With regard to the accelerated reference system K' (space), which requires a simple derivation, the trajectory of light is no longer a straight line (Einstein, 2009). From this it can be concluded that light generally moves in a curvilinear fashion in (superimposed) gravitational fields (Einstein, 2009). This finding is of great significance in two respects (Einstein, 2009).

First of all, a comparison can be made between the movement of light and reality (Einstein, 2009). Even if such an observation leads to the fact that the curvature of light, which is determined by the general theory of relativity, is only surprisingly low with regard to the gravitational fields existing according to experience, then this will actually result in 0.0004722226 degrees with regard to light that shines close to a star such as the sun (Einstein, 2009). This should be verifiable, since fixed stars that can be seen close to a star (the sun), which can correspond to experience through observation when the sun is completely eclipsed, should be seen to be bent by this angle from the star that can be seen the most from the earth (our sun) with regard to their position, which they assume in the earth's planetary sky as soon as the star changes its position in this sky (Einstein, 2009).

This hypothesis test regarding the acceptance or rejection of the described consequence is one of the most important tasks that requires prompt completion by astro-scientists (Einstein, 2009).

According to Albert Einstein (2009), this deflection of light rays predicted by the general theory of relativity during the eclipse of the stars in our solar system as seen from Earth at the end of May in 1919 was captured in a photograph by two astronomers named Eddington and Crommelin, who jointly led expeditions funded by the Royal Society.

Secondly, however, this sequence makes it clear that the law of the constant speed of light c in a vacuum, which is one of two fundamental principles of the special theory of relativity, loses its claim to infinite validity in accordance with the general theory of relativity (Einstein, 2009). The curvature of light only occurs as soon as the speed of movement of the light rays changes due to the place (at which it flies past) (Einstein, 2009). Readers might now think that, on the basis of this sequence, the special theory of relativity, and thus the (entire) theory of relativity as a whole, is disproved (Einstein, 2009). However, this truth (of readers' mental models) is not true (Einstein, 2009). Only one truth is possible, namely that the special theory of relativity requires a finite valid range (in physics), as its results are only valid to a limited extent, so that readers can draw conclusions about natural events (such as the deflection of light) based on the effect of gravitational fields (Einstein, 2009).

Because all critics of the theory of relativity (i.e. the entire general theory including the special theory) often claimed that the special theory is invalidated by the general theory of relativity, Albert Einstein (2009) would like to explain a real fact more comprehensibly by means of a comparison. Before electrodynamics was established, all the laws of electrostatics were accepted (by the critics) as the (most appropriate) laws of electricity (Einstein, 2009). It is now generally understood that the laws of electrostatics can only correctly describe electricity fields in an experiment that has never been realized, so that an electricity mass is actually immobile relative to other electricity masses and with respect to a reference system, i.e. at rest (Einstein, 2009). Did this invalidate the laws of electrostatics using Maxwell's field formulas for electrodynamics (Einstein, 2009)? No (Einstein, 2009). In the laws of electrodynamics, the laws of electrostatics have been retained as a limited area of validity, because electrodynamics leads directly to electrostatics for the state of affairs when the electric field is uninterruptedly constant (Einstein, 2009). The best result of a theory (not only) in physics is if this theory describes the procedure as a basis for the development of a more comprehensive theory (meta-theory), in which the basic theory can continue to exist as a limited area of validity (Einstein, 2009).

Through the (constant) light motion scenario just described, readers could learn that general relativity allows people to understand the effect of the gravitational field on the processes of events in a theoretical way, because for the possibility of the absence of the gravitational field, the laws of light motion have long been understood (Einstein, 2009). However, the most rewarding research task, for which the general theory of relativity provides the method, involves the establishment of laws that describe the gravitational field in a consistent way (Einstein, 2009). The method in the general theory of relativity is as follows (Einstein, 2009).

We know of space-time regions whose behavior is (almost) *galilean* when the reference system is selected in the same way, i.e. (three-dimensional) regions within which no gravitational fields occur (Einstein, 2009). If we now define such an area in relation to a reference system K' that moves at will, then a gravitational field exists with respect to K' that can be changed spatially and temporally, i.e. dynamically (Einstein, 2009). According to Albert Einstein (2009), this can be concluded by abstracting (generalizing) the descriptions contained in section 8. The properties of the gravitational field are of course dependent on the method, i.e. on the selection of the space-time domain motion K' to be determined by us (Einstein, 2009). According to the general theory of relativity, a general gravitational field law determined in accordance with the theory should apply to every gravitational field (Einstein, 2009).

If not every gravitational field can be reproduced by this method, then confidence can still be gained that a general gravitational law can be (successfully) derived with the help of this special class of gravitational fields (Einstein, 2009). This confidence has miraculously not been broken (Einstein, 2009). However, in order to be able to fully understand this goal until it is actually achieved, a decisive barrier had to be overcome first, which Albert Einstein (2009) had to explain to the readers, because it is firmly linked to the matter itself. This again requires a deeper understanding of the concept of the space-time continuum (Einstein, 2009).

11 Properties of scales and clocks on rotating reference systems

Albert Einstein (2009) had deliberately not said a word about the physical understanding of space-time coordinates according to his general theory of relativity. Therefore, Albert Einstein (2009) had blamed himself for allowing a slight incompleteness, because due to the general theory of relativity, readers are aware from the special theory of relativity that space-time coordinates are very important and must be considered. Now is the right time for readers to close the existing knowledge gap by learning; however, Albert Einstein (2009) already said before that there are high requirements for understanding space-time coordinates in terms of readers' nerves and generalization ability.

Readers should again start from frequently quoted, completely unique laws of nature (Einstein, 2009). Readers should imagine a space-time domain (cube) in which no gravitational field exists relative to a reference frame K with a suitably chosen velocity of motion; thus, based on observation of the domain, K represents a Galilean reference frame, and the findings using special relativity are valid relative to K (Einstein, 2009). Readers should imagine the same space-time domain (cube) as a second reference frame K', which has a uniform rotation relative to K (Einstein, 2009). To make the model assumption more detailed, readers should mentally dimension K' as a shape like a flat disk (circle) that rotates uniformly around a center located in its circular surface (Einstein, 2009). Readers (observers) sitting on the outside (eccentrically) of the disk or circle experience an interaction that works eccentrically in the direction of the radius and is perceived by observers as an inertial force (centrifugal effect) relative to the initial reference system K (Einstein, 2009). However, observers (readers) sitting on the circle should understand their circle as an immobile, i.e. resting, reference system K'_0; this is possible for observers on the basis of the general theory of relativity (Einstein, 2009). The influencing effect is understood by (seated, i.e. resting) observers as the force of the gravitational field, which acts on all reference bodies that are immobile relative to their circle (and on them) (Einstein, 2009). Admittedly, the influence of the gravitational field distributed in space represents an interaction that should be excluded according to Newton's theory of gravitation (Einstein, 2009). The gravitational field does not exist in the center of the circle and grows eccentrically (outwards) in proportion to the distance from the center of the circle (Einstein, 2009). However, because observers mentally adhere to the general theory of relativity, they do not mind the deviation from Newton's theory of gravity; they rightly hope that a general theory of gravity could be derived which, in addition to the motion of the fixed stars, also correctly justifies their perceived field of action (Einstein, 2009).

Observers try to set their clocks and find scales on their circle, in the belief that they will obtain exact values for their understanding of space-time coordinates with respect to the reference body K' (disk) on the basis of their perceptions (observations) (Einstein, 2009). What experiences will they have (Einstein, 2009)?

The observers first position two matching clocks, the first in the center of the circle, the second on the boundary line of the circle (periphery), so that the clocks are relative to the reference body K' (disk) or are immobile (Einstein, 2009). First of all, observers should question whether all (two) clocks tick at the same speed from the point of view of the Galilean reference system K, which is moving without rotation (Einstein, 2009). Seen from the reference system K, the clock in the center is motionless without speed, at the same time the clock on the periphery is not at rest due to the rotation relative to K (Einstein, 2009). According to the insight from section 32, the first clock, viewed from K, therefore always ticks faster than the clock on the periphery of the circle (Einstein, 2009). The same result should of course also be confirmed (stated) by observers sitting or resting on the periphery of the circle near the second clock (Einstein, 2009). On that circle and generally within any gravitational field, clocks will therefore tick faster or slower, depending on the position at which the clocks are positioned (stationary or at rest) (Einstein, 2009). An exact value for the time determined by clocks lying or at rest relative to the reference system must therefore be ruled out (Einstein, 2009). Albert Einstein (2009) did not want to go into more detail about the fact that a comparable barrier is revealed as soon as observers attempt to transfer the (previous) concept of simultaneity according to the special theory of relativity to this example.

Concept of simultaneity according to the special theory of relativity (Einstein, 2009). K' rotates uniformly as a circle and moves (in a straight line) with respect to the stationary Galilean reference system K, therefore (further) differently accelerated clocks are possible simultaneously in the center and in the periphery of the circle K' from the point of view of K, thus an exact quantity for time becomes (or and because, according to the special theory of relativity, gravity does not exist as a field, the readers of Albert Einstein (2009) may not imagine their reference system K' to be at rest, which means that gravity exists from the center of the circle and grows eccentrically in proportion to the distance from the center of the circle; Accordingly, Newton's theory of gravitation applies without deviation for this special case and therefore makes it difficult to derive a more general theory of gravitation, because a gravitational deviation only becomes recognizable with the introduced field concept, which is not integrated in the special theory of relativity but necessarily in the general theory of relativity.

However, the determination of the spatial coordinate values also presents observers with insoluble problems for the time being (Einstein, 2009). If, for example, observers moving with the circle place their uniform scale (which is small relative to the radius) touching (tangentially) the periphery of the circle, then the rod, evaluated from the Galilean reference system K, is shorter than 1, since non-resting reference bodies experience a contraction (reduction) in the direction of motion as described in section 32 (Einstein, 2009). If, on the other hand, observers place their unit rod in the straight line of the circle radius, then the rod, evaluated from K, does not experience a contraction (reduction) (Einstein, 2009).

Accordingly, observers resting (sitting) on the circle first measure the circumference a with their unit rod, then the diameter b and then divide (divide) their two measurement results, then the result of the division of a by b (quotient) is a different pi value, instead a higher value, although with their calculation π ($\pi = 3.1415...$) should of course be exactly determinable relative to an immovable circle seen from K (Einstein, 2009).

During the complete operation (calculation), observers must use the non-rotating Galilean reference system K as a reference body (coordinate body), because the validity of the findings according to the special theory of relativity may only be postulated (asserted) relative to K, since a gravitational field exists relative to K' (Einstein, 2009).

This is proof that the laws of geometry according to Euclid are too imprecise for the mathematical observation of a gravitational field, at least as soon as observers try to assign the distance between its two ends (distance) 1 as its x-coordinate (length) to their scale regardless of its position (location) and direction of movement (Einstein, 2009). This also makes it clear to observers that a fixed straight line, which was initially conceptually defined by them as a rod, loses its right to exist (Einstein, 2009). The method used in the special theory of relativity defines that only uniform rectilinear body movements can exist on the basis of the Galilean transformation, which is why it is impossible for observers to precisely (determine) the coordinate values for x, y and z relative to their circle (Einstein, 2009). For this reason, however, space and time coordinates for (all) points (events) must be determined so that natural events, which also contain time coordinates, can also be understood more precisely (Einstein, 2009).

Therefore, it seems that it is now necessary to review every idea that Albert Einstein (2009) had learned so far about the general theory of relativity. In fact, a simpler (more subtle) method is needed so that the basic principle of general relativity can be applied correctly (Einstein, 2009). The readers will be enabled to understand this general principle of relativity with the help of the following descriptions (Einstein, 2009).

12 Is the universe (continuum) Euclidean or non-Euclidean?

Albert Einstein (2009) compared the space-time continuum (our ever faster expanding universe), or rather a selected four-dimensional area of it, with a marble table surface. On this marble table (or in this universe), Albert Einstein (2009) was able to travel from any selected point to any other point by (very) frequently repeating that he would always travel to a neighboring point, or – to put it more accurately – by not skipping any neighboring point, i.e. by not performing or making a jump. Readers will certainly understand with sufficient accuracy (if they do not even have too high an expectation [of Albert Einstein (2009)]) what is meant by neighbor and jump in this context. Albert Einstein (2009) imagined it in such a way that he determined that the surface of the marble table should be understood as a universe (continuum).

Albert Einstein (2009) now imagined a high value for the length and width (dimensions) of the marble table surface (i.e. for the continuum) represented by short, uniformly long rods. This is to be understood in such a way that the ends of the rods are each placed together in line with the next end of the rod (Einstein, 2009). The observers should now place four such rods together on the marble table surface so that four corners are formed from their rod ends and two uniformly long diagonal lines are formed between these corners (i.e. a square is formed as an example of a quadrilateral) (Einstein, 2009). In order to ensure that there is uniformity between the two diagonal lines, Albert Einstein (2009) used a (selected short) rod for testing. Albert Einstein

(2009) added uniform squares to this square, each sharing a (short) rod with the first square, to these four squares and so on. In the end, the entire marble table surface (the entire universe or continuum) is filled with squares, so that all square insides (from left to right and vice versa as well as from top to bottom and vice versa) each belong or point to two further squares and all square insides (left-top, right-top, left-bottom and right-bottom) each belong or point to four further squares.

This task is generally feasible (for every thinker) without encountering the smallest problem, that is (absolutely) true and (therefore) wonderful, thus for true a miracle (Einstein, 2009). In general, only the next thought is needed (Einstein, 2009). If three squares are already adjacent to an inner corner of a square, then two uniform long sides of this fourth square must also already be occupied (Einstein, 2009). How the two remaining uniform long sides of this fourth square are to be occupied is completely determined by the previous thought (for each individual side of the square) (Einstein, 2009). However, Albert Einstein (2009) was by no means able to localize his square as before, i.e. to change its position (location) in the space-time continuum (again), so that its diagonal lines are homogeneous (equal) with regard to their center (to the center of the square) or are of uniform length when viewed side by side (parallel). If both diagonals are already uniform without adjusting the position of the quadrilateral, i.e. a square is present, then this special physical fact is justified with the help of the marble table surface and the (short) rods, which Albert Einstein (2009) could only be grateful for, as this represented a true miracle. Albert Einstein (2009) had to experience a large number of comparable miracles (such as on a marble table surface covered with short rods) if the design of the space-time continuum, i.e. the design of the universe, had to be successfully implemented.

If the marble table surface is actually completely covered with rods that form squares with diagonals (squares) that are uniform on the inside, then Albert Einstein (2009) confirmed that the events (points) on the surface represent a universe (continuum) according to Euclid with regard to the rods used in the form of a path. Albert Einstein

(2009) marked one corner of a square as the starting event (starting point), on which he is able to describe all corners of the other squares with regard to the starting point using two values. Albert Einstein (2009) only had to say how many rods he had to travel in the direction to the right and how many rods he had to travel in the direction above the start event in order to reach his desired destination point, a randomly selected corner of a square. The two characterized values are therefore equal to the Cartesian coordinate values for the number of rods in the right and up directions with respect to the Cartesian coordinate system specified by the staked rods (Einstein, 2009).

There must also be situations (events) that cause the experiment to fail, as Albert Einstein (2009) was able to demonstrate using the next adaptations of this thought experiment. All rods must undergo an expansion (enlargement) in accordance with the laws of temperature (Einstein, 2009). The marble table surface is heated in the center, but not at the (outer) edge, which means that the squares become smaller and smaller from the center outwards and, conversely, larger and larger from the edge inwards, so that, as before, two rods can always be placed in the same position anywhere on the surface (Einstein, 2009). However, the increase in temperature causes chaos (disorder) in the design of the squares, as the rods increase in size within the heated part of the surface, which means that the rods expand in all directions, but the rods in the outer, colder part of the surface remain unchanged (Einstein, 2009).

With regard to the rods – which were (previously) defined as uniform distances (distances) – the marble table surface no longer represents a universe (continuum) according to Euclid, and it was also impossible for Albert Einstein (2009) to define Cartesian coordinate values directly on the basis of the deposited rods, because the previous construction design can no longer be realized as uniform rod squares. However, because other bodies (things) exist that do not react in the same way (or not at all) as the rods due to temperature changes of the surface, it is possible to continue to assume that a universe (continuum) according to Euclid exists for the surface in accordance with general laws of nature; this is more satisfactorily possible with a simpler definition for the measurement or comparison of distances (distances) (Einstein, 2009).

However, if all classes of rods (bodies), regardless of their matter, react equally sensitively to any temperature existing on the differently heated marble table surface, and if Albert Einstein (2009) had no more suitable method of observing this temperature interaction than by the rod behavior according to the geometry in experiments in the same way as in the previously characterized experiment, then it would probably appear to be fundamentally useful to assign the distance 1 to two event points on the surface in each case, if the rod ends can be designed to correspond to each other with the selected test rod; so how could a distance be determined differently (even better) in compliance with the general laws of nature (absence of extreme arbitrariness)? For this reason, however, the previous method, the Cartesian coordinate system, must be discarded and replaced with a more suitable method that can be applied without the requirement for the validity of Euclid's geometry for rigid reference bodies (Einstein, 2009).

According to Albert Einstein (2009), mathematicians had noticed this difficulty according to the next description. If a three-dimensional space measurement was made according to Euclid, for example if the surface of a three-dimensional ellipse (ellipsoid) was determined as a plane, then a geometry existed here in two dimensions that is just as correct as the geometry with respect to the flat area (Einstein, 2009). Gauss solved this difficulty by calculating this surface without exception (in principle) with a two-dimensional geometry, neglecting the fact that this surface corresponds to a three-dimensional space (continuum) according to Euclid (Einstein, 2009). If a construction design with rigid (short) rods is generally transferred or placed on this surface (comparable to the previous square division of the marble table surface), then different theorems are valid for this design compared to the theorems of Euclid's geometry for the surface (Einstein, 2009). This (upper) surface does not represent a space (continuum) according to Euclid with regard to the rods (of different lengths), and no numbers (values) for Cartesian coordinates can be defined in it (Einstein, 2009). Gauss illustrated the general theorems with which the area ratios must be calculated according to their geometry and thus developed a method for calculating multidimensional and non-Euclidean spaces (continua) according to Riemann (Einstein, 2009). This explains why mathematicians have long since eliminated the geometric difficulties that the general principle of relativity (also) entailed (Einstein, 2009).

Readers should understand that the case described in this section is similar to the one created by the general principle of relativity (Section 11) (Einstein, 2009).

13 The Gauss coordinate system

The approach introduced by Gauss of a three-dimensional spatial measurement via the two-dimensional geometric examination is to be understood as follows (Einstein, 2009). In general, an arrangement of individual trajectories (system) can be carved on a marble table surface (Figure 7), which Albert Einstein (2009) called *u-curves* and each of which he labeled (annotated) with a value. Albert Einstein (2009) labeled the trajectories with u = number (Figure 7). Readers should imagine another infinite number of trajectories plotted between the first two trajectories, which correspond to every mathematically possible (real) number between $u_1 < u_2$, for example 1.0000000001 < 1.9999999999 (Einstein, 2009). Accordingly, there exists (in between) a (further) arrangement with *u-curves* (system) that cover the entire marble table surface endlessly adjacent (densely) (Einstein, 2009). Not one of the *u-curves* intersects a neighboring *u-curve*, instead all event points (P) on the marble table surface are intersected by only one (single) *u-curve* (Einstein, 2009).

Each individual point (P) on the marble table surface can therefore be assigned a completely individual *u-number* (Figure 7) (Einstein, 2009). According to the *u-curves*, a *v-curve system* can be plotted on this surface, which fulfills the same conditions (general laws of nature), i.e. has real numbers, but is also individually constructed (Einstein, 2009). Each individual point (P) on the marble table surface can therefore be assigned a *u-number* as well as a *v-number*; Albert Einstein (2009) referred to these two values as the coordinates of the surface (Gauss coordinates).

For example, in Figure 7, point P has the Gauss coordinates v = 1 and u = 3 (Einstein, 2009). The points P and P' as neighboring points on the surface therefore have the Gauss coordinates (32) and (33) (Einstein, 2009).

(32) P(v; u) (Einstein, 2009)

(33) P‘(v + dv; u + du) (Einstein, 2009)

dv and *du* are tiny values of the Gauss coordinates (32) and (33) (Einstein, 2009). The distance between P and P' measured by means of a very short rod would also correspond to a tiny value *ds* (Einstein, 2009). Accordingly, ds^2 is given by Gauss using (34) (Einstein, 2009).

(34) $ds^2 = g_{11}du^2 + 2g_{12}dvdu + g_{22}dv^2$ (Einstein, 2009)

The numbers g_{11}, g_{12} and g_{22} derived from formula (34) depend entirely on *v* and *u* (Einstein, 2009). The numbers g_{11}, g_{12}, g_{22} specify the behavior of the very short rods (light bodies) relative to the *v-curves* and *u-curves*, thus overall relative to the area of the marble table (Einstein, 2009). For the scenario in which the event points (P) of the investigated surface form a universe (continuum) according to Euclid with regard to the very short measuring rods, but also only in this case, it cannot be ruled out to design the *v-curves* and *u-curves* accordingly and to provide them with values, so that the formula (35) applies for ds^2 according to Gauss (Einstein, 2009).

(35) $ds^2 = dv^2 + du^2$ (Einstein, 2009)

If the Gauss formula (35) applies, the *v-curves* and *u-curves* exist [in the universe (continuum)] according to Euclid's geometry in a straight line, which are vertical to each other (Figure 7) (Einstein, 2009). The Gauss coordinates therefore correspond to

Cartesian coordinates (Einstein, 2009). In general, it can be seen that the Gauss coordinates merely represent an annotated arrangement of two values (comments) with regard to the points on a (mathematically) examined surface, so that in space the point neighbors, as of P relative to P', must be as close as possible to the same numbers (Einstein, 2009).

For the time being, the results of the study are only valid for a two-dimensional universe (continuum) (Einstein, 2009). However, this Gauss method can also be applied to three-, four- and multi-dimensional universes (continua) (Einstein, 2009). For example, if a four-dimensional universe (continuum) exists, then the next general consideration (valid and therefore) is given (Einstein, 2009). All event points in the universe are (then) assigned any four values for (the four dimensions) x_1, x_2, x_3 and x_4, which are (also) designated as (Gaussian) coordinates (Einstein, 2009). According to Albert Einstein (2009), x_4 is equal to time t. Point neighbors are coordinate neighbors given in numbers (Einstein, 2009). If a physically correctly determined distance (distance) ds now exists with regard to the point neighbors P and P', as confirmed by repeated measurement, then equation (36) is valid (Einstein, 2009).

(36) $ds^2 = g_{11}dx_1^2 + 2g_{12}dx_1dx_2 \ldots + g_{44}dx_4^2$ (Einstein, 2009)

The numbers g_{11}, g_{12} and g_{44} of formula (36) change with their location in the universe (continuum) (Einstein, 2009). Only in a scenario in which a universe (continuum) according to Euclid exists, it does not appear to be impossible that its coordinate quantities x_1, x_2, x_3 and x_4 can be included accordingly for the event points of the universe, so that formula (37) applies to ds^2 according to Gauss (Einstein, 2009).

(37) $ds^2 = dx_1^2 + dx_2^2 + dx_3^2 + dx_4^2$ (Einstein, 2009)

In a universe (continuum) according to Euclid, expressed with the formula (37), four-dimensional vectors (relationships) are to be understood in the same way as those relationships that are generally accepted as three-dimensional vectors (relationships) when measured (Einstein, 2009).

The derived Gauss equation (37) for the determination of ds^2 appears to be only partially given or correct, actually only in the case where sufficiently small areas of an examined universe can each be regarded as a (separate) continuum according to Euclid (Einstein, 2009). This is obviously true, for example, for the scenario concerning the marble table surface and the differently tempered locations (points) on it (Einstein, 2009). Thus, the temperature existing in a (selected) small area of the surface is to a certain extent constant, because here the geometric behavior of the rods almost corresponds to such a behavior as this is given according to the theorems of geometry according to Euclid (Einstein, 2009). The deviations (errors) in the construction design of the squares therefore have a recognizable effect on relationships (coordinates) for the first time as soon as the design [of the universe (continuum) according to Euclid] grows beyond a large area of the marble table surface (Section 12) (Einstein, 2009).

In summary, Albert Einstein (2009) stated the following theorem: The method established by Gauss for the mathematical consideration of arbitrarily large universes (continua) is based on a determination of distances (rods) between the neighboring locations (points) (physically existing there) or relationships between the measures (coordinates) (mathematically possible there) that applies within these areas (Einstein, 2009). All event points of a universe (area) are accordingly assigned a (matching) number of values (Gauss coordinates), as this universe has (a number of) dimensions (Einstein, 2009). This systematic arrangement of dimensions as Gauss coordinates (values) within the areas (universes) is done accordingly so that a clearly defined or designed dimension system (coordinate system) can be (formed and) maintained, and so that a finite number of equal values (Gauss coordinates) can be added to point neighbors (Einstein, 2009). A Gauss coordinate system (therefore) represents a mathematical abstraction (generalization) of a Cartesian value system for coordinates (Einstein, 2009). The Gauss coordinate system can also be applied to (infinitely large) universes (continua, domains) that *do not behave* dimensionally *Euclidean* [*i.e. also to domains (continua) in which existing arbitrarily defined distances (rods) appear like uniform squares*], but only (*not*) in the case where the dimensionally smaller (*resp. larger*) an observed, mathematically considered sub-region of a universe appears [*i.e. (no longer) from the region (continuum) in which these defined distances (rods) appear like (non-)uniform squares*], as soon as an (*un*)finite dimensional smaller (*resp. larger*) part of the examined (infinitely large) universe (area) shows a stronger (*or weaker*) geometric behavior according to Euclid due to a higher mathematical approximation with regard to the defined distance measure (test rod) (Einstein, 2009).

Based on Albert Einstein (2009), this sentence means in simpler terms: at least four dimensions are sufficient to enable a physically correct understanding of the entire universe (continuum), although our ever faster expanding universe could also have more dimensions – i.e. a greater curvature of space-time – not only in its sub-areas, due to the probably higher differences in gravitational fields that exist there. For example, more dimensions could exist outside (or inside) than inside (or outside) the continuum due to gravitational fields, if there were more (or less) matter outside in the edge region of the universe than inside (and vice versa) (Section 20).

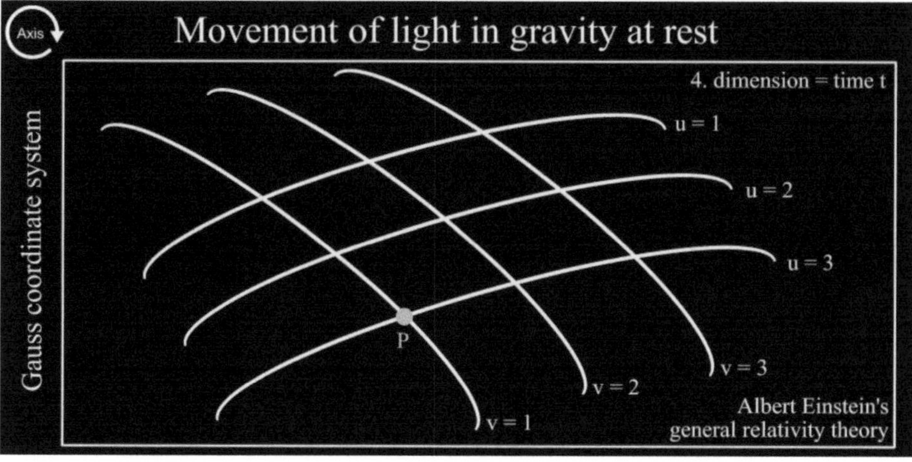

Figure 7. Arrangement of individual path curves (Einstein, 2009, p. 57).

14 The Euclidean universe (continuum) as a space-time continuum (according to Einstein or) as a result of the special relativity theory

Albert Einstein (2009) was now able to express Minkowski's assumption, which was only briefly explained in section 37, a little more precisely (more definitely). For the (mental) construction of the four-dimensional space-time continuum, a certain coordinate system is to be favored according to the special theory of relativity, which Albert Einstein (2009) had called the Galilean coordinate system. In Galilean systems, x, y, z and t are physically simplified as four coordinates that denote a reference body or – in other words – a point event in the universe (continuum) with four dimensions; this is described in detail in the third section of this book (Einstein, 2009). The connection design (of the construction of the universe) between the Galilean coordinate systems moving relatively uniformly to each other is to be understood as the formulas of the transformation according to Lorentz, which represents a basis for the derivation (deduction) of the consequences of the special theory of relativity and even for each Galilean reference body [in the universe (continuum)] only forms the sentence regarding the (everywhere) general validity of the law of the constant speed of light (law of motion of light) (Einstein, 2009).

Minkowski discovered the next simplified assumptions in accordance with the Lorentz transformation (Einstein, 2009). In a universe (continuum) with four dimensions, there are two point neighbors (event neighbors) to be examined, to which a reciprocal location is associated with respect to the Galilean reference system K according to the differences in the space coordinates *dx*, *dy* and *dz* and the difference in the time coordinate *dt* (Einstein, 2009). The same differences exist with regard to the second Galilean reference system K' for the two point events *dx'*, *dy'*, *dz'* and *dt'* (Einstein, 2009). Accordingly, the following relationship (38) is always valid between the two point events (Einstein, 2009).

The derived equations relevant for these coordinates are $x + y + z + \sqrt{-1}ct = x' + y' + z' + \sqrt{-1}ct'$ and $x_1^2 + x_2^2 + x_3^2 + x_4^2 = x'^2_1 + x'^2_2 + x'^2_3 + x'^2_4$; these equations are also valid with respect to coordinate differences, i.e. also with respect to infinitely small coordinate differences (coordinate differentials) (Einstein, 2009).

(38) $dx^2 + dy^2 + dz^2 - c^2dt^2 = dx'^2 + dy'^2 + dz'^2 - c^2dt'^2$ (Einstein, 2009)

The relationship (38) is the reason for the validity of the Lorentz transformation (Einstein, 2009). Albert Einstein (2009) formulated the following theorem: The number ds^2 assigned to both point neighbors in the space-time continuum with four dimensions is the same in every preferred Galilean reference system (39) (Einstein, 2009). Albert Einstein (2009) preferred the left-hand side of equation (39) for the Galilean reference system K.

(39) $ds^2 = dx^2 + dy^2 + dz^2 - c^2dt^2$ *AND* $ds'^2 = dx'^2 + dy'^2 + dz'^2 - c^2dt'^2$ (Einstein, 2009)

If *x*, *y*, *z* and $\sqrt{-1}ct$ are generally exchanged for x_1, x_2, x_3 and x_4, then the general result is also that the number ds^2 is not dependent on the choice of reference system (40) (Einstein, 2009). Here too, Albert Einstein (2009) preferred the left-hand side of equation (40) for the Galilean reference system K.

(40) $ds^2 = dx_1^2 + dx_2^2 + dx_3^2 + dx_4^2$ *AND* $ds'^2 = dx'^2_1 + dx'^2_2 + dx'^2_3 + dx'^2_4$ (Einstein, 2009)

Albert Einstein (2009) described the number *ds* as the distance (path) between the two point events in four dimensions.

The derived constructs are $dx_4 = -1c^2dt^2$ and $dx'_4 = -1c^2dt'^2$ (Einstein, 2009).

If the imaginary constructs $x_4 = \sqrt{-1}ct$ and $x'_4 = \sqrt{-1}ct'$ are generally selected instead of the real constructs for the time t = x_4 and t' = x'_4, then, according to the special theory of relativity, the space-time continuum is to be generally understood (determined) as a universe (continuum) according to Euclid with four dimensions (Section 13) (Einstein, 2009).

15 The non-Euclidean universe (continuum) as a space-time continuum (according to Einstein or) as a result of the general relativity theory

In the third section of this book, Albert Einstein (2009) was able to select the space-time coordinates that allow a physically simple, immediate understanding and that are determined as Cartesian point coordinates with four dimensions (Section 14). Because this choice of coordinates has not been excluded by the law of constant velocity of light propagation, which is however adapted in general relativity (Section 9); Albert Einstein (2009) had come to the more appropriate conclusion that, according to general relativity, the velocity of light motion must always depend on coordinates as soon as a gravitational field (of a reference body) exists. Albert Einstein (2009) also discovered on the basis of a certain thought experiment (Section 11) that the existence of gravitational fields rules out a law-compliant determination of space-time coordinates for the constant speed of light movement, although this was possible in the special theory of relativity (to obtain a law-compliant explanation of the constant speed of light propagation with the same determination of space-time coordinates).

Including these thought results, Albert Einstein (2009) came to a (new, better) certainty (truth) that according to the general theory of relativity the space-time continuum is by no means to be understood according to Euclid, but instead that in the theory of relativity the more general scenario exists, which he had realized for a universe (continuum) with two dimensions for a marble table surface with position-dependent variable temperature. In this two-dimensional universe, it was impossible (due to the local temperature differences) to design a square (Cartesian) coordinate system of uniform rods, so in general relativity it is also impossible to design a coordinate system (reference system) of practically rigid clocks and bodies, so that clocks and scales placed at rest can directly represent time and space relative to other (practically rigid) clocks and scales (Einstein, 2009). This represented the core of the problem that Albert Einstein (2009) solved (Section 11).

However, the descriptions in sections 13 and 14 explain a (theory-based) procedure for working out a solution to the previous problem (Einstein, 2009). To this end, Albert Einstein (2009) described a space-time continuum with four dimensions along an arbitrary path with Gauss coordinates. He annotated all event points in the universe (continuum) with the four values x_1, x_2, x_3 and x_4, which have no direct physical meaning at all, but are instead only intended to number the events in the universe using a fixed but arbitrary procedure (Einstein, 2009). The annotation carried out by Albert Einstein (2009) does not necessarily mean that x_1, x_2 and x_3 represent the three "space" coordinates, x_4 the one "time" coordinate, or that they should be understood as such.

Readers are likely to think that this characterization of their entire environment is completely inadequate (Einstein, 2009). What sense does this make as soon as Albert Einstein (2009) attaches the coordinate numbers x_1, x_2, x_3 and x_4 to the point event, if such variables also have no meaning at all? With a more precise way of thinking, however, readers can imagine that this question itself makes no sense (Einstein, 2009). As an example, Albert Einstein (2009) examined an arbitrarily propagated point event of matter. If the arbitrarily moving point of matter only had an instantaneous, non-persistent constancy, then this point of matter should be represented in space by its individual number system x_1, x_2, x_3 and x_4 (Einstein, 2009). The permanent constancy of the point of matter thus appears to be described by an infinite number of similar number systems, with coordinate numbers that are always related; a point of matter can thus be rectilinear on one dimension in a universe (continuum) with four dimensions (Einstein, 2009). Quite a few (uniformly) accelerated mass points probably also form such straight lines in the universe (continuum) in which everything is located (Einstein, 2009). All possible assumptions regarding the mass points, which have a physical truth or reality, correspond in the universe (continuum) to the assumptions regarding the coincidence of mass points (Einstein, 2009). This coincidence is evident from Albert Einstein's (2009) mathematics-based observation that their two straight lines, which represent their individual points of motion, share the specific arrangement (systematization) x_1, x_2, x_3 and x_4 with coordinate numbers (Einstein, 2009). Since these coincidences really (truly) enable all solely correct assumptions (theories) with a space-time reference, which Albert Einstein (2009) recognized as the content of physical assumptions (theories), observers will confirm this without criticism by means of a thoroughly conducted thought investigation.

When Albert Einstein (2009) previously described the point motion of matter relative to a reference system, he was by no means commenting on something else, such as the coincidence between mass points and defined (fixed) point events with respect to a reference system. Similarly, the reference to time in this respect can be explained by an assumption (theory) about the coincidence between bodies and clocks, linked to the assumption (theory) about the coincidence between clock hands and defined (fixed) straight lines (mass points) on (circular) clock surfaces (clock faces) (Einstein, 2009). The same applies to measurements with a spatial reference using test rods (scales), which require many considerations (by the observer) (Einstein, 2009).

According to Albert Einstein (2009), (the following physical theorization or categorization) is universally valid: All physical theories (assumptions) can be divided into respective sub-numbers of theories (propositions, descriptions), all of which must be related to a coincidence of space with the time of two point events P and P'. All these theories (propositions, contents) can be represented as Gauss coordinates by means of the equality of four coordinate numbers x_1, x_2, x_3 and x_4 (Einstein, 2009). A theory about the space-time continuum described with Gauss coordinates thus really replaces a theory (about the space-time continuum characterized with Cartesian coordinates) completely by means of (the appropriate selection of) the (geometric) reference system, free from the adoption of the weaknesses of Euclidean theorizing as a (selected) method of representation or modelling; this theory does not have to take into account any Euclidean theorem (expression, character) with regard to the universe (continuum) to be modelled and visualized (Einstein, 2009).

16 The exact representation of the general relativity principle

Now it was possible for Albert Einstein (2009) to replace his temporary description of the general principle of relativity in section 6 with an exact representation. His previous (outdated) version, "every reference system such as K and K' etc. is equally suitable for the representation of nature (description of general laws of nature), regardless of the respective current state of motion", can by no means be retained in the same way, since a consideration of practically rigid reference systems in the theorization of space-time according to a description method applied in the special theory of relativity generally makes no sense (Einstein, 2009). The Cartesian coordinate system is (therefore) replaced by the Gauss reference system (Einstein, 2009). A basic understanding of a general principle of relativity is similar to the expression: "*Every Gauss coordinate system is fundamentally equally suitable for the representation of general laws of nature*" (Einstein, 2009).

In general, this general principle of relativity can also be formulated with a further theorem, which makes the same thing possible much more precisely than its nature-coincident extension (expansion) of the special principle of relativity (Einstein, 2009). The laws of general nature in the special theory of relativity are described with formulas that change into formulas of the same kind as soon as new spatial and temporal constructs x', y', z' and t' are implemented (introduced) from another (Galilean) reference system K' instead of spatial and temporal constructs x, y, z and t from a Galilean reference system K by applying the Lorentz transformation (Einstein, 2009). These formulas for describing the general laws of nature in the general theory of relativity, on the other hand, are intended to transfer x_1, x_2, x_3 and x_4 into formulas of the same type by using arbitrary substitutions of Gauss constructs; since every transition (including a transformation according to Lorentz) is equivalent to a transformation from one Gauss reference system into another Gauss coordinate system (Einstein, 2009).

If a lived view (observation) with three dimensions is to be maintained in general, then a (theory-based) evolution, which Albert Einstein (2009) understood on the basis of the well thought-out basic principle of the general theory of relativity, can be described as follows (as a theory development): In the special theory of relativity, Galilean regions exist (in the universe), i.e. only those within which a gravitational field does not exist (Einstein, 2009). The (geometric) reference system selected here is a Galilean reference system, i.e. a practically rigid system with a coincident (matching) selected state of motion, so that the Galilean axiom is valid relative to this system with regard to a rectilinear, uniform direction of motion of "separate" mass points (Einstein, 2009).

Certain ideas make it possible to connect the same Galilean domains to non-Galilean (Cartesian) reference systems as well (Einstein, 2009). A special class of gravitational fields therefore exists relative to this non-Galilean (Cartesian) reference system (Section 8 and Section 11) (Einstein, 2009).

However, practically rigid reference bodies do not exist within gravitational fields and therefore do not have a uniform geometric constitution according to Euclid in this field; this is why the concept of rigid reference systems does not work in general relativity (Einstein, 2009). Gravitational fields also influence the mechanics of clocks, so that compared to the special theory of relativity, the definition of physical time directly determined with clocks justifiably achieves a lower degree of evidence (in the general theory of relativity) (Einstein, 2009). Building on Albert Einstein (2009), the marble table surface and the flame, which led to a heat-induced deformation (contraction) of the square construction design of the universe (continuum) according to Euclid, should be recalled at this point.

Non-rigid reference systems are therefore generally used, which move arbitrarily overall and also experience arbitrary changes in their construction design during this flow of movement (Einstein, 2009). For the determination of time, clocks are used with arbitrary, strongly deviating non-uniform mechanical laws, which are generally mentally fixed (at rest) at every point of a non-rigid reference system, and which have only a single task (determination), to display the measurements (values) of local clock neighbors that can be observed at the same time infinitely in agreement with each other (Einstein, 2009). The same non-rigid reference system, which could generally be called a "reference soft body" for good reason, corresponds for the most part to nothing more comparable than an arbitrary Gauss coordinate system with four dimensions (Einstein, 2009). The difference between a "reference soft body" and a Gauss coordinate system (reference fixed body) consists of a certain simplicity, the (to a certain extent unfounded) lawful retention of the peculiarity (deformation) of spatial coordinates that show a unique property (anomaly) compared to the temporal coordinate (Einstein, 2009). While the reference soft body is considered as a reference system, all point events of the reference soft body are considered as spatial coordinates, all mass points that are immobile relative to the reference soft body are considered as mass points at rest without exception (Einstein, 2009). The general principle of relativity is similar to the requirement that every reference soft body can be used as a reference system for the description of generally valid laws of nature on the basis of the same claim and the same result; these laws of nature must be able to exist completely independently of a selection of the reference soft body (Einstein, 2009).

All laws of nature are assigned an extensive restriction (limitation) due to their independence of the reference soft body selection, which is similar to the discovery effect (explanatory power), which is (also) integrated in the general principle of relativity (Einstein, 2009).

17 The general relativity principle as a solution to the problem of gravitation

If the readers were able to understand all the previous thoughts, then they will (also) be able to understand the approaches (methods) for solving the problem of gravitation without any further problems (Einstein, 2009).

Albert Einstein (2009) started the investigation from a Galilean region, i.e. one in which there is no gravitational field relative to the Galilean reference system K. According to the special theory of relativity, the (physical) behavior of clock mechanics and test rods (scales) with respect to K is predetermined, as is the behavior of "separate" point masses; which are uniformly rectilinear in motion (Einstein, 2009).

As a next step, Albert Einstein (2009) linked this Galilean domain to an arbitrary Gauss coordinate system or to a "reference soft body" as reference system K'. A gravitational field G (special class) therefore exists with respect to K' (Einstein, 2009). The behavior of the clock mechanics and test rods (scales) as well as separately movable point masses with respect to K' is generally determined by means of transformation alone (mathematical conversion of coordinates) (Einstein, 2009). This (physical) behavior is generally understood as the behavior of the clock mechanisms, test rods and point masses under the influence of a gravitational field G (Einstein, 2009). The assumption is then generally introduced that the influence of a gravitational field on the clock mechanisms, test rods and separately movable point masses also follows the same laws of nature in the event that an existing gravitational field *cannot* be derived from the Galilean-specific scenario using pure coordinate conversion (Einstein, 2009).

As a subsequent step, the space-time behavior of the gravitational field G obtained by means of a Galilean-specific scenario using pure point conversion and deduction (derivation) is generally considered and this behavior is described with a law of nature that applies continuously, also regardless of the selection method of the reference system (reference soft body) used for its formulation (Einstein, 2009).

This law of nature does not yet represent a *general* law of nature about the gravitational field, because this investigated gravitational field G represents a special class (Einstein, 2009). For the discovery of a generally valid gravitational field law, a generalization (abstraction) of the correspondingly derived law of nature must also take place, but this can be systematically discovered if the three following conditions are met (Einstein, 2009):

(A) The generalization (abstraction) to be achieved must also satisfy the general principle of relativity (Einstein, 2009).

(B) If an (arbitrary) form of matter exists in the investigated area, then only its mass inertia, i.e. only its mass energy, should be responsible (or causal) with regard to its gravitational field-generating interaction (Section 35) (Einstein, 2009).

(C) (The field of) matter (energy) as well as (its) gravitational field are both obliged to fulfill the law of conservation of energy (as well as momentum) (Einstein, 2009).

Eureka the general principle of relativity enabled Albert Einstein (2009) to determine the effect of a gravitational field with regard to the occurrence of each of these events, which happen in the scenario of a missing gravitational field according to derived laws of nature, i.e. are already contained in the quadrangle of the special theory of relativity. In general, the method developed with regard to clock mechanics, test rods and separately movable point masses is used as a guideline (Einstein, 2009).

A gravitational field theory extracted as described by derivation (deduction) of the general principle of relativity is also characterized by its clarity (brilliance), this also removes (eliminates) the weakness mentioned in section 9, which sticks to Newtonian mechanics, this also enables an explanation of the law of experience regarding the equivalence between mass inertia and mass gravity, this also already offers two different interpretations of fundamentally different results of observational astronomy, with regard to which Newtonian mechanics fails (Einstein, 2009). In the first observational result, Albert Einstein (2009) referred to the orbit of a planet called Mercury; in the second astronomical result, which he had already mentioned, he referred to the bending of light due to the gravitational field of a star (the deflection of light after the sun caused by its gravitational field).

If the formulas for the general theory of relativity are generally concentrated (specialized) on one scenario, so that all gravitational fields do not behave strongly, and that each point mass moves with a speed against the (selected) reference system that is low compared to the speed of the light rays, then the result is generally first in the form of a first approximation of Newton's (known) theory; This is therefore determined in the general theory of relativity without a special hypothesis, although Newton was forced to integrate the interaction of gravitational attraction, which is indirectly proportional to the square of the distance between the point masses, as (an additional) assumption (Einstein, 2009). If the precision of the operation (calculation) is generally increased, then differences with regard to Newton's theory are noticeable, which, however, due to their minimal numerical property, have been forced to escape experience (observation) almost altogether until now (Einstein, 2009).

Albert Einstein (2009) had to emphasize one of the differences in the general theory of relativity. According to Newton's theory, planets move in ellipses around a star such as the sun; these ellipses would leave their location unchanged forever with respect to the fixed stars if the gravitational interaction of existing planets on a planetary mass to be investigated and the own motion of fixed stars were negligible (Einstein, 2009). If these two effects are neglected, an orbit of a planetary mass must represent an ellipse that remains constant against fixed stars if Newton's theory exists exactly correctly (Einstein, 2009). For every planet, with the exception of the planet Mercury, which orbits closest to the Sun on an ellipse, this sequence, which can be demonstrated with remarkable accuracy, was confirmed with a clarity that makes observational clarity achievable today (Einstein, 2009). However, Albert Einstein (2009) knew from Leverrier with regard to the planet Mercury that its ellipse of the orbit corrected as described above is not unchanged in comparison with its fixed stars, but instead circles, naturally extremely leisurely, in its orbital plane equal to its orbit of motion (Einstein, 2009). With regard to the movement of the rotation of the elliptical orbit, an angle of 0.0119444 degrees was calculated every hundred years, the angle of which is rounded to a few degrees (Einstein, 2009). A justification of this phenomenon according to Newton's theory is only possible with the aid of assumptions that are entirely conceived for this reason and are hardly possible (Einstein, 2009).

Based on the general theory of relativity, it is confirmed that all elliptical orbits of the planets must rotate around a star like the sun as described, that the rotational motion for each planet – except Mercury – is too small to be confirmed with the observational acuity achievable today, but that in the case of Mercury this has to be 0.0119444 degrees every hundred years, exactly as confirmed by experience (of astronomy through observation) (Einstein, 2009).

In addition, only one further consequence has since been derived from the general theory of relativity, which allows verification by experience (observation), namely the spectral shift (red shift) of light rays sent here not from small but from giant stars compared with the light rays produced on planet Earth using the same process (i.e. by means of the same class of particles) (Einstein, 2009). Albert Einstein (2009) firmly believed in his prognosis (prediction) that the last consequence of his general theory of relativity would also be confirmed promptly; as he assumed, this was proven close in time by observation.

Views on the entire universe

18 Limitations of Newton's theory with regard to the universe

In addition to the limitation (restriction) described in section 9, there is another fundamental limitation in Newton's theory which, to the best of Albert Einstein's knowledge (2009), was first criticized in detail by an astronomer named Seeliger. As soon as a general assumption becomes necessary that describes the entire universe as a thought, for example, then this probably corresponds directly to the following formulation (Einstein, 2009). The universe has infinite space (including time) (Einstein, 2009). Stars exist everywhere, so the density of matter is of course extraordinarily different in several areas, but it corresponds to an equal density in the sense of an extended average in every place (in every area) (Einstein, 2009). In other words, in general, regardless of how deep into the space of the universe one goes, there in every place travelers will discover an approximately equal number of fixed stars of approximately the same class and matter density (Einstein, 2009).

The latter choice of words cannot be reconciled with Newtonian mechanics (Einstein, 2009). Instead, Newton's theory has the condition that the universe has a certain class of a spatial center, where the greatest density of stars is found, which decreases from the inside to the edge of the universe, so that an infinite matter-free space can exist away from the center outside in the edge area (Einstein, 2009). All stars in the universe would have to form a locally non-infinite (four-dimensional) land area in a never-finite (four-dimensional) water area in space (Einstein, 2009).

Albert Einstein (2009) sees the reason for this: According to Newton's theory, a set of "lines of action" in a point mass m have their one end, whose other end arises from infinity, and whose quantity is calculated in proportion to the point mass m (Einstein, 2009). If the mass density p_0 in the universe is equal on average, then this point mass is located on average in a sphere with the volume V, which leads to the mass density p_0V (Einstein, 2009). A total quantity of lines of action, which are connected from the surface of the sphere F to its center, thus corresponds proportionally to p_0V (Einstein, 2009). For the quantity of lines of action on the surface of the sphere (all of which are connected to a center), the result is equal to p_0V/F or p_0R (Einstein, 2009). The surface concentration of the lines of action or their field intensity on the surface of the sphere should therefore expand to infinity as the radius R increases, which appears to be impossible (Einstein, 2009).

This idea (regarding the origin of the formation of matter in the universe) hardly seems to be satisfactory (based on Newton's theory) (Einstein, 2009). The idea appears to be all the more improbable, since it can generally be concluded that every ray of light and its countless emitting stars travel separately in the (entire) stellar space (universe) without interruption into infinity, i.e. never return and never again establish a physical interaction with other matter objects existing in the stellar nature (Einstein, 2009). A universe in the center of which a non-infinitely condensed mass m has accumulated should therefore evaporate proportionally (again) piece by piece (Einstein, 2009).

To prevent this possibility (consequence) from becoming a reality of the universe, Seeliger adapted Newton's theory accordingly so that gravitation between two point masses m decreases in the case of large distances r greater than the original Newton's theorem m/r^2 (Einstein, 2009). As a result, it is possible for the average density of matter at any location in the universe to remain the same towards infinity, thereby preventing a non-finite field intensity of the gravitational fields (Einstein, 2009). Accordingly, this unnecessary idea that a universe consisting of matter must have a certain class of center is generally forgotten (Einstein, 2009). Of course, this forgetting generally takes place on the basis of the fundamental requirements described by means of no adaptation or increase in the complexity of Newton's theory of gravitation based on observation (experience) and theory (Einstein, 2009). Therefore, all arbitrarily conceivable theories can explain the same thing (i.e. gravitation) equally, although no justification would be possible for this, so that one of the theories could be better suited than one of the other theories; since, just as with Newton's theory of gravitation, each of these theories lacks any justification with the help of generalized theory-based principles (Einstein, 2009).

19 The theory of a finite and yet infinite universe

However, other existing hypotheses regarding the structure of the universe also grew in a completely different theoretical direction (Einstein, 2009). For example, the emergence of a geometry existing without reference to Euclid led to a (new) truth that a skeptical opinion about the existence of an infinite space is generally justified without contradicting any of the thought models or models of experience (Einstein, 2009). Albert Einstein (2009) refers to Helmholtz and Poincaré, who corrected these facts in detail with unsurpassed ingenuity, which is why he only explained them briefly here.

Albert Einstein (2009) first envisioned an event (happening) on two dimensions. Two-dimensional beings have two-dimensional methods, mainly two-dimensional rigid scales (test rods), and are movable separately (from each other) on a flat domain (Einstein, 2009). Far from the two-dimensional boundaries of this flat domain, there would be no (other) reality for these beings, instead it is an event (happening) within their flat domain that they personally experience (observe) individually and on their plane-based objects, (which therefore) (expresses) a closed causality (Einstein, 2009). Above all, the construction design of a flat geometry according to Euclid appears to be realizable on the basis of the rods (of the beings), for example the already investigated net-shaped arrangement (system) with squares on a marble table surface (Section 12) (Einstein, 2009). The universe of creatures appears to be a space with two dimensions in contrast to the universe of Albert Einstein (2009), but like his 2 + n dimensioned universe it continues to infinity. (Sufficient) space has an infinite number of uniform squares, each consisting of four rods (due to the infinity of the universe corresponding to the 2 + n dimensions), therefore the volume (surface) related to these squares is equal to infinity (Einstein, 2009). For these creatures it has a meaning to express, my universe presents itself "flat", for example the meaning that their construction design with the rods is possible due to the geometry according to Euclid, whereby each rod always symbolizes the same distance, not depending on its location (space) (Einstein, 2009).

Albert Einstein (2009) now imagined an event (happening) in two dimensions for the second time, but no longer on the surface, instead on the surface of the sphere. All two-dimensional beings, including their test rods and other objects, are located exactly on the surface of the sphere and are not able to disappear from this universe; the entire universe in the form (geometry) of the perception (based on experience) of the beings is instead extended to this surface of their sphere without exception (Einstein, 2009). Is it possible for the beings to judge a geometry (form) of their universe with the help of Euclid's geometry on two dimensions and meanwhile their own test rods (scales) as a realization of a "distance" (in accordance with reality) (Einstein, 2009)? The beings have no way of doing this (Einstein, 2009). Since in their experiment to make their line straight, the beings instead always get a curved or crooked line (geodesic), which Albert Einstein (2009) called "beings on three dimensions" with the widest disk, thus a straight line closed (connected) at its beginning and end with a fixed non-infinite size, which can be measured with a rod length (Einstein, 2009). In the same way, the universe of beings has a non-infinite surface, which is comparable to a square area made up of four rods (Einstein, 2009). The best cognition that enables such thoughts to be deepened consists of the stimulus (Einstein, 2009): *The universe of three-dimensional beings does not correspond to infinity as well as nevertheless never possesses finiteness* (Einstein, 2009).

However, all spherical beings do not have to undertake an expedition through their universe in order to realize that the universe in which these beings live is not Euclidean (Einstein, 2009). In all areas of their universe, which are very large, it is possible for the three-dimensional beings to check the non-Euclidean geometry of their universe (Einstein, 2009). Starting from a point in any direction, the beings draw "non-odd distances" (viewed in three dimensions, disk arc lines) of matching size (Einstein, 2009). The connection of exposed end points of the drawn distances is called "disk" by the beings (Einstein, 2009). The division of a disk circumference verified using the test rod by the radius diameter measured using the same rod results in the invariant π within a surface according to Euclid's geometry, which has no dependence on the disk diameter (Einstein, 2009). The beings conceived by Albert Einstein (2009) should discover the size (41) on their spherical surface for this division.

(41) $\pi \frac{\sin\left(\frac{r}{R}\right)}{\left(\frac{r}{R}\right)}$ (Einstein, 2009)

The expression (41) represents a quantity that is smaller than π and, of course, the smaller the higher the disk radius is compared to the spherical radius R of the "spherical universe" (Einstein, 2009). By means of this ratio, it is possible for the spherical beings to determine the radius R of their universe, even if they only have a comparatively small area of their spherical universe available for their measurements (Einstein, 2009). If, however, their area turns out not to be large enough, then these beings have lost their ability to notice (state) whether they are located in a spherical universe and not in a circular universe according to Euclid; because the component of their spherical surface that is measured as too small is very similar to a correspondingly extended component of their flat area (Einstein, 2009).

Therefore, if these spherical beings live on top of some planet, and this planet is located in a star system that only fills the (entire) spherical universe as an inconspicuously minimal area, then the beings have no knowledge that helps them to determine whether the universe they inhabit is finite or infinite, because the area of the universe that they can experience through observation is approximately flat or Euclidean according to these two scenarios (Einstein, 2009). This idea by Albert Einstein (2009) directly conveys that, with regard to their spherical being, a disk circumference first increases in terms of radius until the "universe circumference" is reached, so that its disk circumference then shrinks back to zero piece by piece (rod by rod) as the radius grows ever larger. In the meantime, a disk plane constantly increases in size until it finally coincides with the entire plane of its entire spherical universe (Einstein, 2009).

Readers may wonder why they should think of the spherical surface and no other non-open surface for the placement of their beings (Einstein, 2009). However, this way of thinking has a justification according to Albert Einstein (2009), since this spherical surface has been endowed with a function (by nature) compared to the other non-open planes, making every point on it indiscriminate. The relationship of a circumference u of a disk with the corresponding radius r does not appear to be independent with respect to r, but in the case of a certain r it appears to be the same with respect to every point in the spherical universe; the spherical universe presents itself as a "plane of constant curvature" (Einstein, 2009).

With regard to the spherical universe with two dimensions, there is an analogy with three dimensions, the spherical (curved) space with three dimensions, which Riemann found (Einstein, 2009). Every point in curved (spherical) space is also synonymous (Einstein, 2009). The curvature of space (sphere of space) does not have an infinite volume of space, which can be defined by $2\pi^2R^3$ using the radius of the sphere R (Einstein, 2009). Is a sphere of space (space curvature) generally conceivable (Einstein, 2009)? The idea of a space means to mentally experience an image of a space-related observation (experience), i.e. on the basis of experiences that can be carried out (in nature), which can generally arise through the movement of practically rigid reference bodies (Einstein, 2009). With this way of thinking, a spatial sphere (spatial curvature) can be experienced (Einstein, 2009).

Starting from an (arbitrary) point, Albert Einstein (2009) drew lines (connected ropes) in every direction and measured the same distance r with respect to each (straight) line using his test rod (scale). Each exposed end point of the (uniform) distances marks a spherical surface F (Einstein, 2009). Albert Einstein (2009) determined a plane of F by precisely measuring a square formed with his test rod. If a universe exists according to Euclid, then $4\pi r^2$ equals F; if the universe corresponds to a sphere, then $4\pi r^2$ is always greater than F (Einstein, 2009). F increases together with increasing r from zero to the maximum scale (maximum test scale) conditionally by means of the "universe radius", until F shrinks back to zero rod by rod (piece by piece) with steadily increasing radius r of the sphere (Einstein, 2009). Starting from its starting point, all (drawn) radial lines first drift further and further apart until they subsequently return so that they finally connect back at the end point (counter-event) of the starting point; the straight radial lines are thus circled through the entire sphere of space (Einstein, 2009). This quickly leads to the general conviction (theory) that a spherical plane with two dimensions is to be understood completely the same (analogously) as a spatial sphere with three dimensions (Einstein, 2009). Space does not exist infinitely (i.e. with no infinite spatial volume), with missing barriers (boundaries) (Einstein, 2009).

It should be noted that there is another intrinsic class of the space sphere, the "space ellipsoid" (Einstein, 2009). The space ellipsoid is to be understood as a space sphere in which all "counter-events" exist equally (indiscriminately) (Einstein, 2009). A universe in the form of a space ellipsoid is therefore to be experienced geometrically as a universe that has a non-asymmetrical space sphere in the center (and that has an increasingly folded elliptical space on the outside due to further superimposed dimensions of centrally non-asymmetrical space spheres) (Einstein, 2009).

Based on the statements (theories), it can be concluded that (according to logic, by no means only) a non-open space with missing barriers (boundaries) exists (Einstein, 2009). Conceivable spaces under consideration are (in particular) the space sphere (or the space ellipsoid) due to clarity (naturalness), which becomes clear because all associated event points exist with the same meaning (Einstein, 2009). As a result of these statements, an astonishingly serious question arises in the minds of (not only) all physicists and astronomers: does the universe, within which everyone lives, not exist finitely or, according to the assumption of a sphere of space for this universe, not exist infinitely (Einstein, 2009)? For the counter-thought (answer) to this thought (question), the experience gained so far is by no means sufficient in the slightest (Einstein, 2009). However, the general theory of relativity enables a counter-thought (answer) to this thought (question) that is determined with a high degree of accuracy; this also removes a limitation (restriction) described in section 18 (Einstein, 2009).

20 The spatial structure that exists in the general relativity theory

There is no geometric self-determination of space according to the general theory of relativity, but there are functions of space due to its geometry determined by mass (Einstein, 2009). In general, based on this, only a conclusion about an (existing) geometry-based model of the universe (spatial structure) is possible, under the condition that the status of mass cannot be included as an unknown in the investigation (Einstein, 2009). Albert Einstein (2009) knew from his observation (experience) the physical fact that the speed of movement of light is large compared to the speeds of stars when a reference system (coordinate system) is chosen to match. Albert Einstein (2009) was therefore able to understand the (existing) model components of the universe (spatial structures) essentially through the coarsest (mathematical) approximation by imagining each (moving) mass as an unmoving (resting) mass.

Albert Einstein (2009) already knew from previous trains of thought that their clock mechanics and test rods (scales) are influenced in their behavior by gravitational fields, i.e. by the propagation of mass. From this it can already be concluded that a geometry according to Euclid is not possible with exact (consistent) validity in the universe as a theorem (Einstein, 2009). However, the following seems conceivable in principle, namely that (only) a small deviation (difference) exists between the geometry according to Euclid and the universe, which, understood as an assumption (theory), is all the more probable, since this equation confirms that even matter with the volume of the sun has only a completely negligible influence on these model properties (model metrics) of its surrounding space (Einstein, 2009). In general, it would be conceivable for the spherical beings that their universe is constructed according to the same geometry as a plane that is asymmetrically curved in many places, but which is not strikingly different at any point compared to the straight surface, in contrast to, for example, a water surface folded by means of low waves (Einstein, 2009). Albert Einstein (2009) agrees that such a universe would be described as an approximate Euclidean universe. The space of this universe would not be finite (Einstein, 2009). However, the equation confirms that within an approximate Euclidean universe, the average mass density would be zero (Einstein, 2009). A corresponding universe would therefore only be populated in some places full of mass; this universe would resemble the same unsatisfactory model of spatial structure that Albert Einstein (2009) developed in section 18.

However, if the model must have a spatial structure like the universe, i.e. an average mass density that is as low as possible with respect to zero, then this universe does not correspond to any approximate truth according to Euclid (Einstein, 2009). Instead, the equation confirms that the universe should exist as a space sphere (or space ellipsoid) conditioned by the model scenario of uniformly distributed mass (Einstein, 2009). Because this mass is in reality asymmetrically dispersed in many places, the true universe must be different from the behavior of a space sphere at many points, this universe should correspond to an approximate space sphere (Einstein, 2009). However, due to the model, this universe should not be allowed to exist infinitely (Einstein, 2009). Furthermore, the general theory of relativity provides an easily comprehensible chain of thought regarding the universe modeled by its space expansion in relation to its average mass density (Einstein, 2009).

The formula determined for this with regard to the "radius" R of the universe is $R^2 = 2/kp$ (Einstein, 2009). If the CGS system of units is used, $2/k$ equals 1.08×10^{27}; where p represents the average mass density (Einstein, 2009).

Third section: On the special relativity theory

21 Axiom about the relative change in position of practically rigid bodies

The readers have certainly already become acquainted with the extensive structure of geometry according to Euclid and probably remember this extensive structure with respect rather than enthusiasm, the difficult hurdles of which have been drilled into them by accurate lecturers during countless lectures (Einstein, 2009). Based on this experience, the readers would certainly be averse to anyone who would say that even the slightest axiom of geometry is not true (Einstein, 2009). However, this emotional thought of comprehensive certainty should probably quickly leave the readers if Albert Einstein (2009) were to ask them: "What is meant by this hypothesis, so that the axioms cannot be untrue?" Albert Einstein (2009) wanted to take a closer look at this question.

Geometry contains certain basic understandings, for example surface, circle, line, on the basis of which Albert Einstein (2009) was able to link less or predominantly unambiguous views, as well as with regard to certain short propositions (axioms), which he had to tend to accept as "not untrue" because of these views. The other axioms must then be justified, i.e. proven, with regard to these propositions because of the method of logic, the confirmation of which Albert Einstein (2009) had conditionally accepted. The axiom only appears correct or "not untrue" if it is derived according to the accepted method on the basis of the propositions (Einstein, 2009). A question regarding the "correctness" of the respective geometric axioms therefore leads back to the question regarding the "correctness" of the theorems (Einstein, 2009). However, it has long been certain that there is no answer to this question using any method, not even the geometric method, i.e. that this question makes no sense at all (Einstein, 2009). In general, it is not even possible to ask whether it should not be untrue that only one (single) line runs through two points (Einstein, 2009). In general, it can only be said that Euclid's geometry consists of shapes which this "line" designates and to which it assigns the function of being described without error on the basis of two corresponding points (Einstein, 2009). The understanding "not untrue" in no way agrees with the expressions of absolute geometry, since Albert Einstein (2009) was always used to designating this straight line in accordance with the "real" object on the basis of the designation "not untrue"; a (Euclidean) geometry, however, does not deal with the relation of associated understandings with regard to objects of experience, but only with a logical linking of the understandings with each other.

It is easy to understand why Albert Einstein (2009) nevertheless tended to regard all geometric axioms as "not untrue". All understandings of geometry are predominantly or hardly correctly similar to objects in our natural environment, the perception of which is of course the only reason for the evolution of these understandings (Einstein, 2009). Geometry deviates from nature in order to attach the best possible closed logic to its structure; a habit, namely to perceive two marked positions within a straight line with respect to a practically rigid body, is deeply hidden in our habitual ways of thinking (Einstein, 2009). The habit of Albert Einstein (2009) was to additionally think of three places localized on a line if he was able to experience their observation points (places) in a summarily straight line in one direction apparently by means of concordant selection of the place of perception while seeing with only one eye.

Albert Einstein (2009) added only one theorem (or axiom) to Euclid's theorems (axioms) of geometry due to this habitual way of thinking, because this already makes the correspondingly supplemented Euclidean geometry an important model component of physics. The axiom added by Albert Einstein (2009) about the relative change in position of practically rigid bodies is: *The distance between two points on a practically rigid body is always the same, i.e. the distance, regardless of the change in position* (Figure 8). With this reasoning, it is now possible to ask about the "correctness" of correspondingly understood geometry theorems, as this allows us to ask whether these axioms are fixed with regard to the natural objects that Albert Einstein (2009) thought were attributed to all understandings of geometry. Thus, Albert Einstein (2009) was not able to express very precisely that by the term "correctness" of geometry theorems he meant their agreement during a design with circular tools as well as line tools.

Any certainty regarding the "correctness" of geometry theorems is, of course, only due to very incomplete experiences (Einstein, 2009). Albert Einstein (2009) initially assumed this correctness of all geometry theorems in order to finally recognize within the second section of these investigations (on the general theory of relativity) that and in what way this correctness actually has limitations. Therefore, Albert Einstein (2009) first used Euclid's geometry in his special theory of relativity to represent uniform rectilinear motion, and then in his general theory of relativity with a non-Euclidean geometry to point out that the classical geometry imagined by humans generally does not correspond to the nature of gravitational fields, which causes non-uniform curvilinear motion; For example, there are no uniform straight lines, so in an advanced geometry a triangle, for example, must consist of three outwardly curved straight lines (geodesics), and accordingly all other forms of perception, i.e. geometry, must undergo an adjustment that is consistent with nature.

A practically rigid body is a body occurring in nature to which a straight line from geometry is mentally associated (Einstein, 2009). Albert Einstein (2009) cites a triangle consisting of three points A, B and C as an example of a practically rigid body. Point B is drawn between points A and C in such a way that a straight line is created between the three points A, B and C (Einstein, 2009). This means that point B is located between points A and C, which makes the sum of the distances $\overline{AB}$ and $\overline{BC}$ as small as possible, i.e. minimal (Einstein, 2009). This sum is minimal, for example, in an isosceles triangle between points A and C exactly in the middle, where point B must therefore also be given.

Physics corresponds to a model of reality and therefore contains a truth about the modelled reality (Einstein, 2009), for example about the universe and the bodies in it, which are contained in the model and can therefore be visualized (mapped). In this context, the words practical refer to approximate and rigid to unchangeable or immovable, therefore reference is made to an approximately unchangeable or immovable body (Einstein, 2009). Such a body is also represented in nature by a sphere, for example a planet like our earth. A straight line can also be assigned to this sphere, whereby this straight line becomes a geodesic due to the given spherical shape (spherical geometry). The added geodesic in this spherical geometry does not change even if the sphere moves. The direction of the movement is irrelevant. It follows from this: The movement, regardless of the direction, does not lead to any change in the position of points and geodesics on an approximately unchanging or unmoving body. The points A, B and C on this body remain unchanged and thus also the respective distance (distance) between these points. In a space such as the universe, therefore, approximately invariant or immobile bodies can exist even though they are always moving.

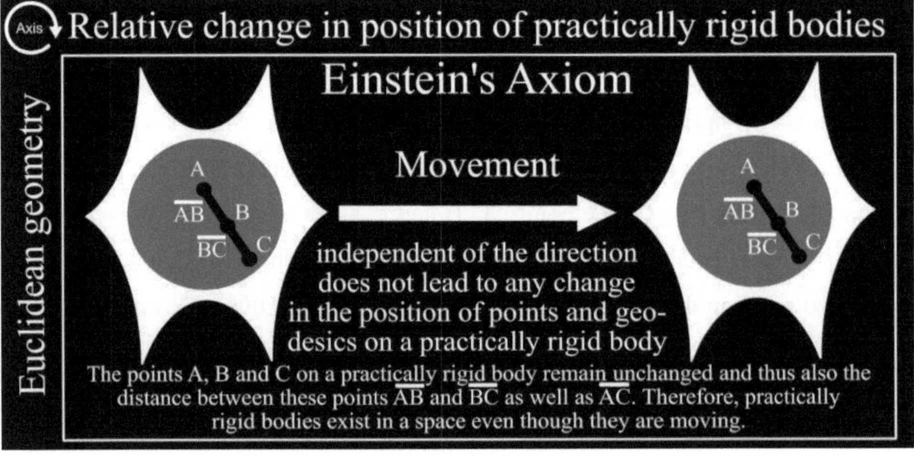

Figure 8. Euclidean geometry and relative position change of practically rigid bodies.

22 The Cartesian coordinate system (of measuring physics)

Due to a circumscribed interpretation according to physics with regard to the distance, Albert Einstein (2009) was also able to specify the distance between two points on a (practically) rigid geometric body by counting (measuring). For this, Albert Einstein (2009) required a distance (bar K) that could never be changed and that should be used as a uniform scale. If A and B now appear as two point markings of a (practically) rigid reference body, then the linking line appears to be configurable according to the laws of geometry (geometry assumptions); Albert Einstein (2009) was then able to place the distance K on the linking line, starting from A, correspondingly often until B was generally reached. The number of repetitions of the distance corresponds to the scale number (measurement) of the distance $\overline{AB}$ (Einstein, 2009). Every measurement of distance is based on this (Einstein, 2009). However, the idea here is that a measurement number would (always) add up, i.e. (always) result in a complete number (Einstein, 2009). This hurdle is generally overcome by using split scales (test scales), the explanation of which requires the same handling (method) in principle (Einstein, 2009).

All spatial descriptions of positions (locations) of an experience or object are based on the fact that a point is generally specified by a (practically) rigid geometric body (reference body) with which the experience (point event) corresponds (coincides) (Einstein, 2009). Due to its validity, this can be used for academic and everyday descriptions (Einstein, 2009). If Albert Einstein (2009, p. 4) examines the positional description "in Berlin, on Potsdamer Platz", then the following means the same: The (practically) rigid body corresponds to the ground on the planet, here Earth, to which a layer designation is assigned, on the body "Potsdamer Platz in Berlin" appears as a labeled geometry point linked to naming, with which the experience in space coincides (coincides). A further analysis of what "spatial coincidence" means at this point seems unnecessary, since this understanding seems understandable insofar as, within a separate natural scenario, differences of interpretation as to whether the term is true or not should not occur often (Einstein, 2009).

The previous simple class of layer labels only allows for layers on the rigid body surface and appears to be dependent on the existence of different surface points

(Einstein, 2009). Albert Einstein (2009) learned by watching how his human mind stood out with respect to these two limitations (constrictions), but without the sense of layer labels undergoing any modification. If there is a weightless thundercloud "above Potsdamer Platz", then the position of this cloud relative to the ground (body surface) is fixed by the fact that a connecting rod is generally erected vertically on this square, which extends upwards to this cloud (Einstein, 2009, p. 4). A distance of the connecting rod, measured using a uniform scale and linked to a layer designation of the lower end of this connecting rod, thus corresponds to a complete layer designation (Einstein, 2009). On the basis of this example, Albert Einstein (2009) recognized the extent to which a detailed (improved) understanding of the layer was achieved.

A) In general, a rigid reference body to which a layer designation is assigned is extended using this method so that the object to be detected can be hit based on the completed rigid reference body (Einstein, 2009). In this case, the cloud becomes a star (Einstein, 2009).

B) In general, a *number* is used to describe the position instead of designated points of reference (in this case, a distance of the connecting rod measured using a test scale) (Einstein, 2009).

C) In general, the distance of the object cloud (or star) is also spoken of in the case where the connecting rod that ends at the object does not exist at all (Einstein, 2009). In this example, it is generally understood from optical photographs of the storm cloud (or star) from different points on the surface, taking into account the movement function of the light, what length this connecting rod should have so that the object can be reached (Einstein, 2009).

Based on this idea, it is generally recognized that for the characterization of positions it is better, once this is achieved, to design unconditionally by means of the use of dimensional numbers regarding the presence of reference points associated with designations on a (practically) rigid reference body to which a position designation is assigned (Einstein, 2009). Our measuring physics achieves this by using a system with Cartesian coordinates (Einstein, 2009).

The Cartesian coordinate system of measuring physics (Figure 9) is also described by Albert Einstein (2009) as a practically rigid body consisting of three mutually vertical and rigid, flat outer walls. The point of an event can be described (not only) by the lengths of the three coordinates x, y, and z with respect to the (practically rigid) Cartesian coordinate system (Einstein, 2009). The lengths can be seen from the event to these three flat outer walls (Einstein, 2009). By means of sequences (counting) of fixed, i.e. rigid rods (units of measurement) or manipulations of such rigid rods (measures), these lengths can be found with respect to the three coordinates x, y, and z (Einstein, 2009). (Possible) manipulations with rigid rods are given by the geometry, laws and methods of Euclid (Einstein, 2009).

The coordinates x, y, and z cannot be determined directly or not really (Einstein, 2009). Coordinates can only be determined indirectly through the use and design of rigid rods (measures, units of measurement), whereby the rigid outer walls of the Cartesian coordinate system are often not realized, which means flexible (Einstein, 2009). Clear, meaning precise, are results of (theoretical) physics, especially quantum physics and astrophysics, when the physical meaning of the point coordinates (x, y, z) are examined according to the previous explanations (Einstein, 2009). The general theory of relativity, which is explained in the second section of this book (Einstein, 2009), requires a more detailed and adapted version of this opinion.

It should now be clear that a practically rigid body is a central prerequisite for the spatial modelling and visualization of events in order to connect these events spatially, i.e. relativistically (Einstein, 2009). This connection has the condition that the geometry according to Euclid's laws is also applicable to distances (distances), whereby in measuring physics the physical distance is always composed of two points located on a practically rigid body (Einstein, 2009).

The Cartesian coordinate system as a practically rigid body represents an infinitely large cube for the supersymmetric and relativistic modelling and visualization of space and time. For example, coordinates (x, y, z) can be determined and shifted to the coordinates (x', y', z') by shifting another coordinate system (K') relative to its original coordinate system (K) instead of the spatial location information itself. This displacement of a coordinate system K' with the axes x', y', and z' relative to its original coordinate system K with the axes x, y, and z reveals the fourth dimension, that of time t to t'. Figure 9 shows an example of such a movement relative to each other, i.e. relative movement starting from the z axis to the z' axis, where zz' is equal to tt'. The speed v of the movement can be determined by integrating this fourth dimension of time t to t' on the respective coordinate system axis. The best way to determine the velocity v of the movement is to assign a rigid coordinate system to each practically rigid body. Any number of practically rigid coordinate systems can be used to model and visualize this movement.

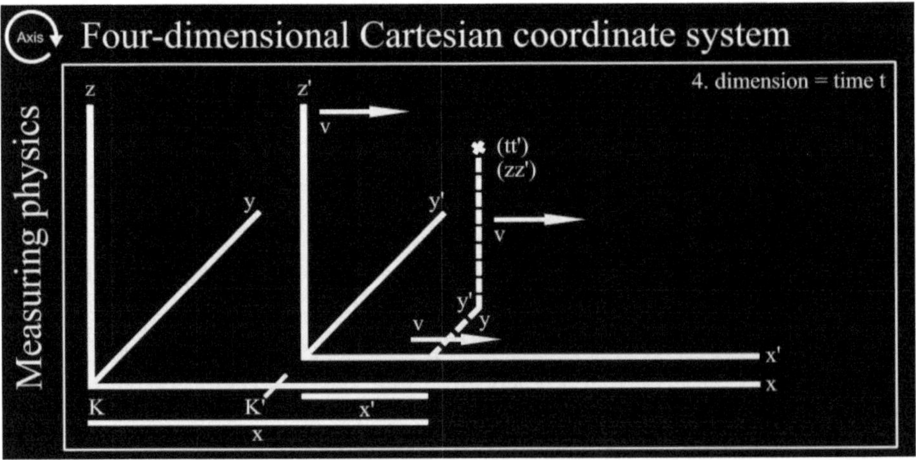

Figure 9. The Cartesian coordinate system of measuring physics (Einstein, 2009, p. 22).

23 Space and time in Newtonian mechanics

Albert Einstein (2009) said the following about classical motion (Newtonian mechanics) without giving any negative thoughts or detailed explanations. For Albert Einstein (2009), the goal of Newtonian mechanics (Figure 10) was: "This theory of motion must characterize all bodies that vary their position in space as a function of time". Albert Einstein (2009) thus conscientiously (or deliberately) accepted several serious errors, even though they violated his clear *ability to cognize*, i.e. contradicted his mind, which was open to the cosmos; the previous errors are shown first.

This does not seem clear, i.e. what can be meant by "spaces" and "positions" in this section (Einstein, 2009). Albert Einstein (2009) stands at a window glass of a uniformly (homogeneously) accelerated spatial reference body E and sends a beam of light to the planetary surface of the Earth, but without redirecting additional (kinetic) energy to it. Albert Einstein (2009) then sees (neglecting the reciprocal effect of atmospheric drag) his beam of light falling downwards as a straight line. The observers who see this process from the planetary surface of the Earth understand that the beam of light radiates down to the bottom of the planet as a parabolic bend (arc shape) (Einstein, 2009). Albert Einstein (2009) now asked: Are the "layers" that our light beam passes "in physical reality" on the line or arc shape? How should spatial mechanics be understood in addition (Einstein, 2009)? One consideration appears to be clear according to the answers in section 22 (Einstein, 2009). First of all, Albert Einstein (2009) had to completely omit the term "space", which still seemed dark, i.e. not bright, because, if he was honest, he could not imagine anything under it; he thought of "mechanics with respect to an approximately resting relation body" as a substitute. All positions in relation to the body (space reference body or planetary surface) were already described (or delimited) in detail in section 22. Because Albert Einstein (2009) established a suitable word "reference system" instead of "reference body" with regard to the mathematical coordinate view, he was able to express the following: A ray of light characterizes a relative line starting from a spatial reference body to which a coordinate reference system is firmly (rigidly) assigned, an arc shape viewed from a coordinate reference system firmly (rigidly) placed on the planetary surface. In general, readers should clearly understand from this example that no orbital curve can actually exist (in our solar system and therefore in the entire universe), which is a bend on which reference bodies move, but only an orbital curve relative to a selected coordinate system (reference body) (Einstein, 2009).

For the movement of a body, a moving body describes a straight line with respect to a coordinate system rigidly harmonized with location A as space A and a parabola with respect to a coordinate system rigidly harmonized with location B as space B (Einstein, 2009). A specific coordinate system is chosen as a specific reference body, therefore there is no trajectory for an observer at location A and another observer at location B in which a specific body can move (Einstein, 2009).

However, the complete representation of mechanics does not appear to be realized until it is generally commented on how a reference body changes its position depending on time, i.e. with regard to all points on the curve, it must be indicated when the reference body is located at this point (Einstein, 2009). The previous annotation must be completed by means of a corresponding definition of the concept of time, so that the time quantities are to be regarded as perceptible units (results of the measurement counts) with the aid of this definition (Einstein, 2009). Albert Einstein (2009) complied with this condition – without abandoning Newtonian motion – with regard to his example using the following method. Albert Einstein (2009) imagined two exactly matching clock-hand machines; Alber Einstein holds one of them in his hand at the space reference window, while the observers on Earth look at the second clock. Everyone notices in which (space-time) area of the respective reference system the beam of light is currently located with each tick of the clock hand, which they hold in their fist. Here Albert Einstein (2009) neglected to explain the lack of accuracy that arises due to the (assumed) finiteness (infinity) of the speed of movement of the light beam. This and the other limitation (barrier) that predominates in this section will be discussed in detail later (Einstein, 2009).

Albert Einstein (2009) thus completed this representation of the movement of bodies by specifying that a body changes its location in space over time and that it must therefore be possible to see for all points on the respective trajectory (straight line and parabola) when a body is at each point in time or when a body will move there or was there (Einstein, 2009). This definition leads to the completion of values over time, so that these data can be considered as generally observable measurement results (units of magnitude) (Einstein, 2009). Albert Einstein (2009) therefore introduced an example according to Newtonian mechanics. He gave the two observers, one of whom is at location A and one at location B, each an exactly technically identical clock (Einstein, 2009), such as a clock hand. Now each of the two observers can determine (at any time) where the body is on the respective coordinate system (reference body) each time the clock ticks (Einstein, 2009). Since bodies change their position in space with time (Figure 10), one observer sees the body falling downwards from above with each rigid unit of time elapsed on the clock in a straight line (Einstein, 2009). The other observer, on the other hand, sees the sideways movement of the falling body from top to bottom on the parabola (Einstein, 2009). For both observers there is therefore a different trajectory in which the body moves with each unit of time, yet both observers see how the body reaches the next reference body (coordinate system) in accordance with their clocks (Einstein, 2009).

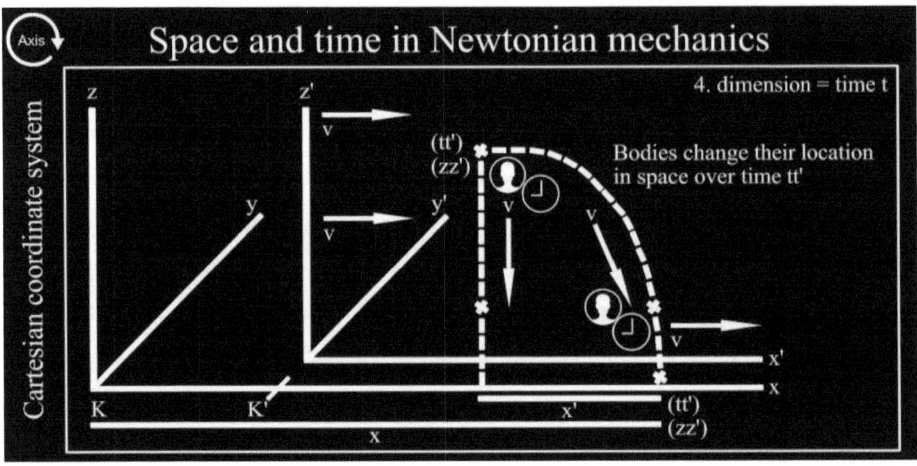

Figure 10. Space and time in Newtonian mechanics (Einstein, 2009, p. 22).

24 The coordinate system in Galileo-Newtonian mechanics

The basis of Galileo-Newtonian mechanics corresponds to the law of inertia (Figure 11) (Einstein, 2009): "*Every body that is sufficiently separated from other bodies remains within a state of rest or a linear-homogeneous (straight-line uniform) celestial mechanics*". This law of inertia is used in the mechanical representation of bodies and their motion and describes suitable reference bodies (coordinate systems) (Einstein, 2009). An example of bodies to which this law of inertia applies with a high degree of credibility are fixed stars, which must be visible so that a coordinate system can be assigned to each of these stars as a reference body (Einstein, 2009). If a planet, such as the Earth, is also rigidly assigned a coordinate system, this leads to fixed stars forming a circle with a huge radius in the course of astronomical days, i.e. following a uniform circular motion, which contradicts the law of inertia (Einstein, 2009). From an observer's point of view, an astronomical day describes the length of time until a star of a planet can be seen again at the highest point in the sky of the same planet. The duration of an astronomical day depends on the speed of rotation of the planet in relation to the fixed stars visible from there (Einstein, 2009). As long as this law of inertia is not to be rejected, movements of bodies may only be related to coordinate systems to which fixed stars do not perform uniform circular but uniform rectilinear movements relatively, i.e. comparatively (Einstein, 2009). The Galilean coordinate system enables movement on a reference body (coordinate system) relative to another reference body (coordinate system) in accordance with the law of inertia (Einstein, 2009). The laws of Galilean-Newtonian mechanics only apply to one of the two Galilean coordinate systems (Einstein, 2009).

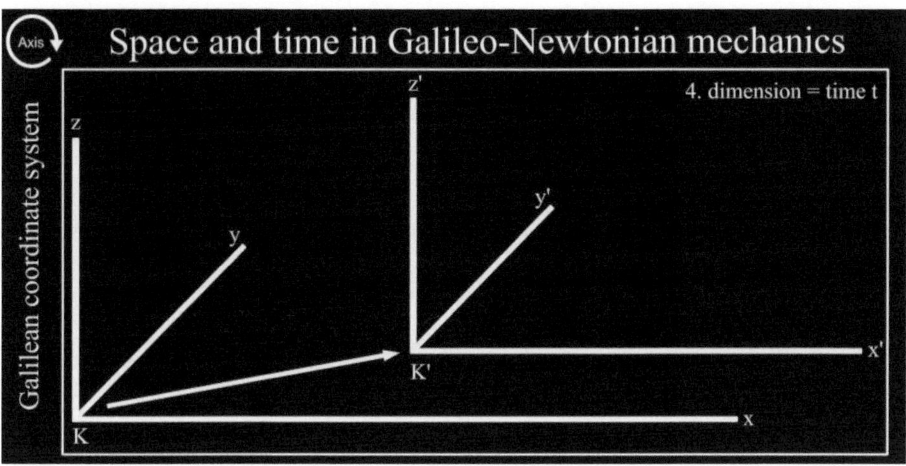

Figure 11. The Galilean coordinate system (Einstein, 2009, p. 22).

25 The relativity principle (within the actual sense)

The mechanics (motion) of a uniformly moving, i.e. accelerated, body is called uniform translation (Einstein, 2009). Uniform refers to a constant, i.e. constant speed in one direction and translation means that the body changes its position relative to the reference body (coordinate system) without rotating (Einstein, 2009). If a body B now flies uniformly and in a straight line – viewed from location B – through space, then viewed from the accelerated body A – the movement of body B is a movement with different speed and different direction, but this is just as uniform and straight (Einstein, 2009). Generally formulated (Einstein, 2009): Thus, if a mass m is moved uniformly and rectilinearly mentally in accordance with a coordinate system K, then this movement takes place just as uniformly and rectilinearly with respect to another coordinate system K', if K' performs a uniform translation with respect to K. This results in a sequence taking into account the description of the previous section (Einstein, 2009):

If K is a Galilean coordinate system, then every other coordinate system K' is also such a system that performs a uniform translation with respect to K (Einstein, 2009). The laws of Galilean-Newtonian mechanics apply to both coordinate systems (Einstein, 2009).

According to Albert Einstein (2009), the principle of relativity (in its true sense) is as follows (Figure 12): If K' with respect to K moves uniformly and without rotation, i.e. is rotation-free, then nature behaves with respect to K' according to exactly the same and general laws of nature with respect to K (Einstein, 2009).

The principle of relativity applies as long as all Galilean coordinate systems, such as K, K', K", etc., move relatively uniformly in relation to one another, i.e. are equally suitable for describing the laws of nature (Figure 12) (Einstein, 2009). If this principle of relativity did not apply, then the laws of nature could only be described if one of the Galilean coordinate systems was designated as K_0 with a specific translation as the reference body (Einstein, 2009). Accordingly, all other Galilean coordinate systems K would be in motion and the coordinate system K_0 would be the only completely stationary system because it is better suited to describing nature (Einstein, 2009). For example, if the location B were the coordinate system K_0, then a body would be the coordinate system K with respect to which more difficult laws should apply than with respect to K_0 (Einstein, 2009). This difficulty would exist because the body K would (actually) be in motion with respect to K_0 (Einstein, 2009). For the general laws of nature described with regard to K, it would have to be assumed that the size and direction of the body's velocity are important (Einstein, 2009). If a planet, such as the earth, is chosen as the body, which moves on a circular orbit around its star, such as the sun, at a speed v of 30 km/s, it would have to be assumed, if the principle of relativity were false, that the current spatial direction of the movement of the planet (earth) is to be regarded as a law of nature, which means that the behavior, i.e. the direction of movement, of all other physical reference systems would have to be dependent on the spatial direction of movement of the planet (earth) (Einstein, 2009). Due to the regular time values, i.e. the annual change in direction of the velocity v of the movement of the planet (Earth) on its orbit, this planet (Earth) cannot exist continuously at rest relative to the hypothetical reference system K_0 (Einstein, 2009). According to Albert Einstein (2009), such an anisotropy (directional dependence) of the physical planetary (earthbound) space, i.e. a physical difference, i.e. a lack of equivalence, of the different directions, has never been observed. For Albert Einstein (2009), this is a detailed justification for the validity of the principle of relativity.

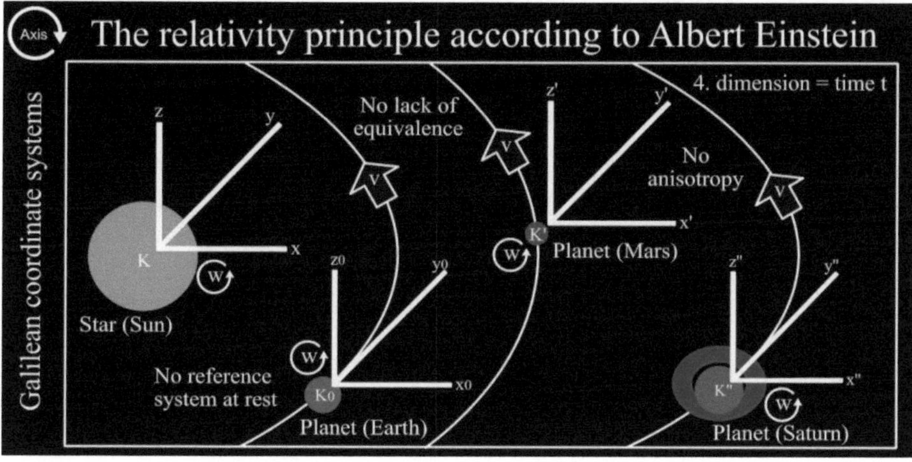

Figure 12. The relativity principle according to Albert Einstein using the example of arbitrarily selected celestial bodies (Einstein, 2009, p. 22).

26 The law of addition of velocities in Newtonian mechanics

A body moves along its trajectory with a constant velocity v (Figure 13) (Einstein, 2009). Another body moves on this reference body K in its longitudinal direction, i.e.

with velocity w in the same direction of motion (Einstein, 2009). What is the velocity W of the second body relative to the first body during the forward motion (Einstein, 2009)?

If the second body is immobile for a unit of time, for example one second, then its forward movement relative to the reference body K' would be a distance v that is equal to the speed of movement of the reference body K (Einstein, 2009). Since the second body is actually moving, it also traverses a distance w relative to the reference body K – and therefore also relative to the reference body K' – in this unit of time (second) through its movement, which corresponds to the speed of its movement (Einstein, 2009). The distance traveled in the assumed unit of time (second) by the second body relative to the reference body K corresponds to (42) (Einstein, 2009).

(42) $W = v + w$ (Einstein, 2009)

In a later section, it becomes clear that the idea of the law of addition of speeds in Newtonian mechanics cannot be maintained and that this law is therefore not correct in reality (Einstein, 2009). Until then, Albert Einstein (2009) wanted to assume this correctness (truth).

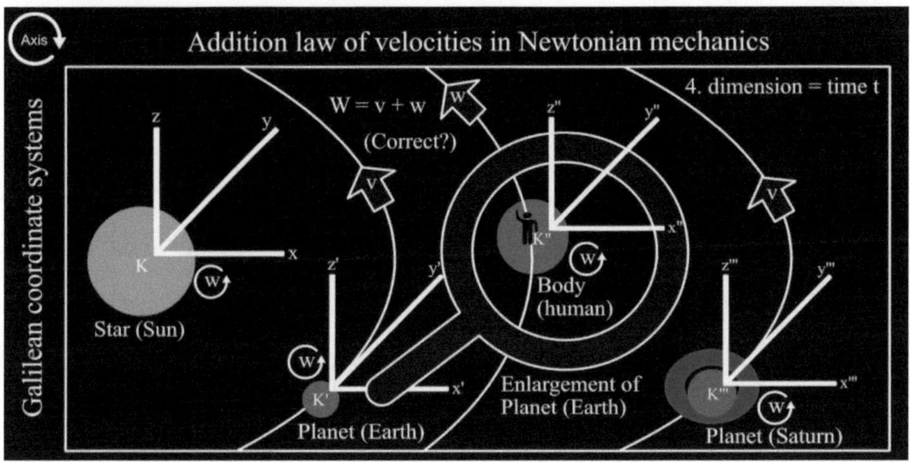

Figure 13. Addition law of velocities in Newtonian mechanics using the example of arbitrarily selected celestial bodies (Einstein, 2009, p. 22).

27 The alleged incompatibility between the law of motion of the speed of light and the relativity principle

The law of the constant speed of light c (Figure 14) is one of the simplest laws in physics (Einstein, 2009). The speed c at which light moves in a straight line in a vacuum, i.e. in empty space without an atmosphere, is 300,000 km/s (Einstein, 2009). This speed of movement of light cannot depend on the speed of movement of a light-emitting body (Einstein, 2009). In addition, it is likely that the speed of light in space does not depend on the direction (Einstein, 2009).

Like any other body, the movement of light (Figure 14) must be related to a rigid reference system (coordinate system) (Einstein, 2009). Starting from location A in a vacuum, a light beam is sent along this location, which moves relative to location A at the speed of light c (Einstein, 2009). A body with the slower speed v moves in the same direction as the light (Einstein, 2009). What is interesting is the speed of movement of light relative to this body (Einstein, 2009). The law of addition of the velocities of Newtonian mechanics (Section 26) can be applied because the light beam moves relative to the body (Einstein, 2009). (43) applies to the speed of light w relative to location A (Einstein, 2009).

(43) $w = c - v$ (Einstein, 2009)

The speed of movement of the light (Figure 14) relative to the body is therefore not c but less than c (Einstein, 2009). This result does not correspond to the principle of relativity (Section 25) (Einstein, 2009).

According to the principle of relativity, the law of the constant, i.e. constant, speed of light c in a vacuum (Figure 14), like all other general laws of nature, should apply to the body as well as to light, each considered as a reference system (Einstein, 2009). As this contradicts the principle of relativity, this result appears to be impossible (Einstein, 2009). The law of the speed of light appears to be a different law of nature with regard to the body, as light moves at the speed c with regard to location A (Einstein, 2009).

This dilemma leads to the conclusion that either the law of the speed of light or relativity (Figure 14) should be rejected (Einstein, 2009). Albert Einstein (2009) regarded the law of the constant speed of light c in a vacuum as a given due to its connection to Lorentz's theory of electromagnetic processes. Therefore, Albert Einstein (2009) could not derive a more complex, i.e. improved, law on the speed of light that was compatible with the principle of relativity. For Albert Einstein (2009), this was the decisive reason why theoretical physicists tended to reject the principle of relativity rather than the law of the speed of light.

The theory of relativity aimed to put an end to this dilemma (Einstein, 2009). By means of an investigation of space and time, it became apparent that there is no incompatibility between the speed of light and relativity (Figure 14) and therefore, by retaining both laws, the result is a comprehensible, contradiction-free theory (Einstein, 2009). Albert Einstein (2009) referred to this theory, which served as the foundation for his extension, as the special theory of relativity, which is described in the following sections by the basic assumptions included in it.

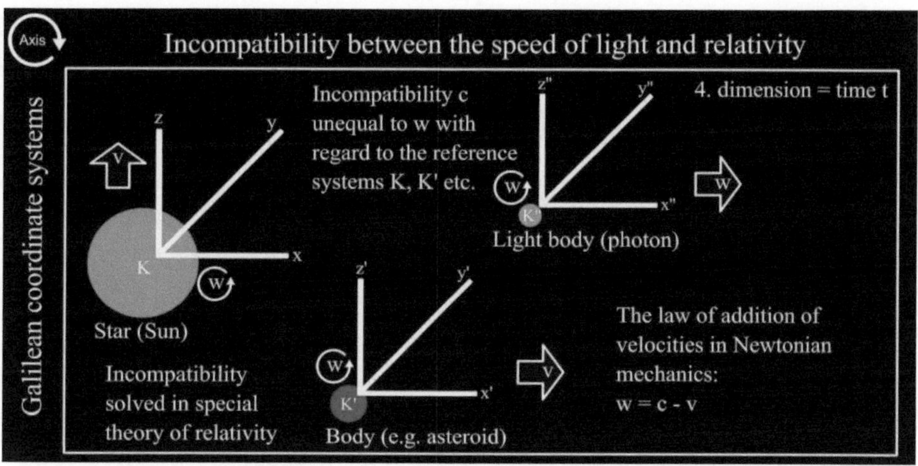

Figure 14. Incompatibility between the speed of light and relativity (Einstein, 2009, p. 22).

28 The concept of time in physics

Albert Einstein (2009) approached the concept of time in physics by explaining the simultaneity of an infinite number of events in an infinite number of places (Figure 15). For Albert Einstein (2009), time as such did not appear to exist; for him, time was merely a quantity associated with the event. For the derivation of the definition of time via the law of the constant speed of light c, please refer to the original work by Albert Einstein (2009) (Reference section). The definition of time according to Albert Einstein (2009), i.e. simultaneity, is as follows: Simultaneity is only possible if corresponding clock hand positions are synchronized at the same speed, which is the case if the clocks attached motionlessly at different locations with regard to a reference system (coordinate system) are coordinated with each other so that changes in the hand positions can occur synchronously at the same speed. In simpler terms, this means that all clocks must function synchronously at an accelerated rate, otherwise this classic definition of simultaneity according to Albert Einstein (2009) does not apply.

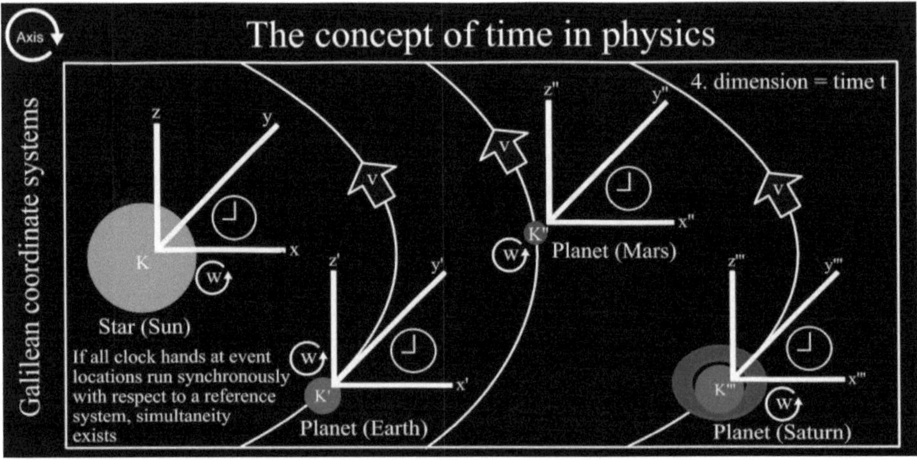

Figure 15. Definition of time in terms of simultaneity using the example of arbitrarily selected celestial bodies (Einstein, 2009, p. 22).

29 Simultaneity or about the relativity of time

Albert Einstein (2009) wrote about the relativity of time (Figure 16), i.e. simultaneity, that events that are asynchronous with respect to one location (reference system) are synchronous with respect to another location (reference system) and vice versa. Each reference system (coordinate system) has its own time in the form of its assigned clock hand position, which means that annotated time values must always refer to a reference system (Einstein, 2009).

Time (Figure 16) is not absolute time as was assumed in Newtonian mechanics, but time is dependent on motion and is therefore not immobile, which means that time is not absolute (Einstein, 2009). If the assumption of absolute time is rejected, there is compatibility between the law of the constant speed of light c and the principle of relativity (Section 27) (Einstein, 2009).

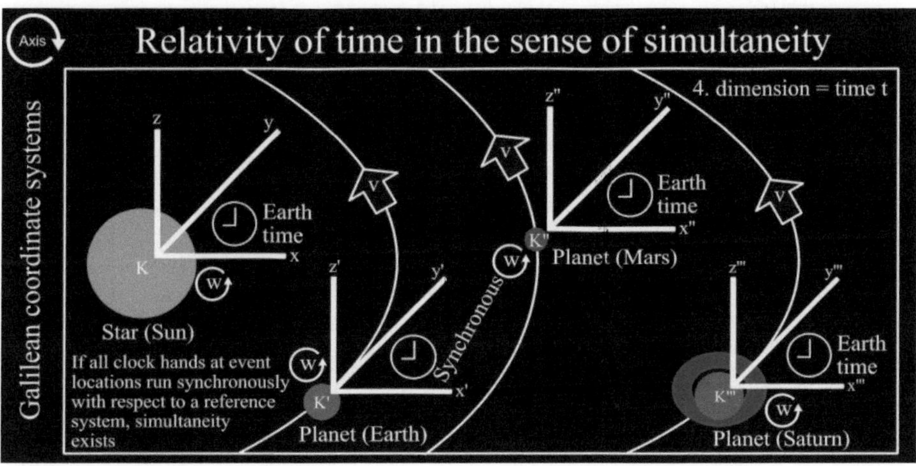

Figure 16. Relativity of time in the sense of simultaneity using the example of arbitrarily selected celestial bodies (Einstein, 2009, p. 22).

30 On the relativity of distance in space

The distance traveled in space does not have to be the same (Figure 17) (Einstein, 2009). This results from the definition of time and the relativity of simultaneity (Sections 28 and 29) (Einstein, 2009). If a body moves on a reference body (coordinate system) and covers the distance (distance) w, then this distance observed from another reference body (coordinate system), which simultaneously covers the distance v in the sense of synchronous clock hands, does not also have to be the same distance as w (Einstein, 2009). According to Albert Einstein (2009), the relativity of distance existing in space applies to bodies moving with respect to another reference system, whereby v does not have to correspond to w. For each light body (photon) with respect to another reference light body (coordinate system), based on the relativity of simultaneity and distance according to Albert Einstein (2009) for the law of the constant speed of light c, it applies that v is always equal to w, i.e. c, if the fixed clock of the second photon at rest is set synchronously with respect to the first light body (Figure 17).

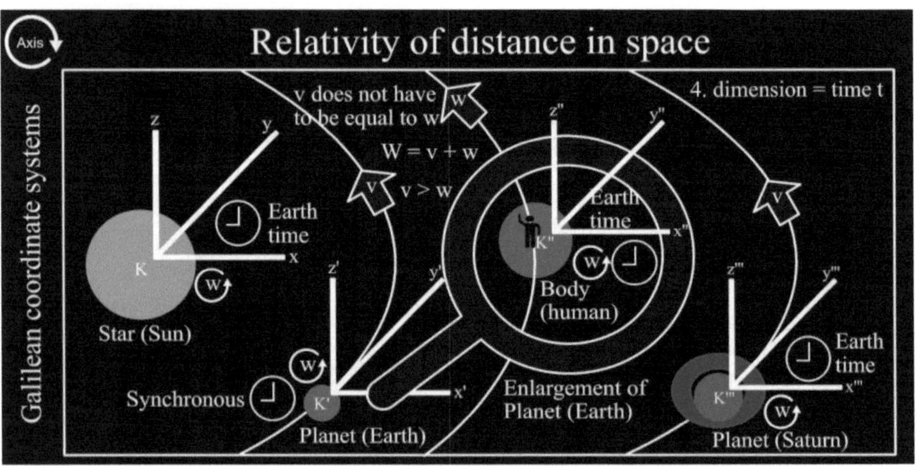

Figure 17. Relativity of distance in space using the example of arbitrarily selected celestial bodies (Einstein, 2009, p. 22).

31 The transformation according to Lorentz

The compatibility of the law of the constant speed of light c with the principle of relativity is due to the fact that two hypotheses from Newtonian mechanics are no longer valid, i.e. have been rejected by Albert Einstein (2009) (Section 27). These assumptions are (Einstein, 2009): (1) The duration of time between two points does not depend on the motion of a reference system and (2) the same applies to the spatial distance between two events of an immobile body.

The dilemma explained in section 27 is thus solved, as the law of addition of velocities presented in section 26 is no longer valid (Einstein, 2009). According to Albert Einstein (2009), there is therefore a solution to the question of whether it is possible that the law of the constant speed of light c does not contradict the principle of relativity? There is a relationship between the time and location of individual points with regard to both reference frames, such that the light has the speed of movement c relative to the location and relative to the body. The solution consists of a transformation law for the values of an event in space-time when moving from one reference system to another (Einstein, 2009).

Albert Einstein (2009) established the two coordinate systems K and K' for two reference systems on the basis of the solid body theory, which states that solid bodies are impenetrable. An event with respect to K is spatially defined by the three axes x, y, and z for the point surfaces and temporally by a quantity t for the time; correspondingly, the same spatio-temporal fixation for the same event takes place for K' according to physical measurement results (44) (Einstein, 2009).

(44) K‘(x‘, y‘, z‘, t‘) ≠ K(x, y, z, t) (Einstein, 2009)

The difficulty from (44) is (Einstein, 2009): If the magnitude fixations for K are given, what are these magnitudes for K' with respect to the identical event? The selection of the relationships for K and K' for an identical body of light (photon) is made according to the law of the constant speed of light c (Einstein, 2009). This difficulty is resolved for the relative and spatial alignment of the coordinate systems K and K' modeled and visualized in Figure 9 using formulas (45) to (48) (Einstein, 2009):

(45) $x' = \frac{x-vt}{\sqrt{1-\frac{v^2}{c^2}}}$ (Einstein, 2009)

(46) $y' = y$ (Einstein, 2009)

(47) $z' = z$ (Einstein, 2009)

(48) $t' = \frac{t-\frac{v}{c^2}x}{\sqrt{1-\frac{v^2}{c^2}}}$ (Einstein, 2009)

This system of formulas is known as the Lorentz transformation (Einstein, 2009). Using the Lorentz transformation, Albert Einstein (2009) achieved the same result as in section 30 for the evaluation, i.e. confirmation, of the constant c contained in the law of the constant speed of light.

For a comprehensible mathematical derivation of this Lorentz transformation, please refer to the original work by Albert Einstein (2009) in the appendix (Section Reference).

The Lorentz transformation shows that the law of the constant speed of light c in a vacuum applies to the reference systems K and K' (Einstein, 2009). This is shown by the fact that a ray of light moves along the x-axis from zero to the positive side at the speed c according to the formula (49) (Einstein, 2009).

(49) $x = ct$ (Einstein, 2009)

According to (44) of the Lorentz transformation, there is a dependency between x and t as well as x' and t' (Einstein, 2009). If the term ct is used for x in (45) and (48), Albert Einstein (2009) obtained (50) and (51) for the first and fourth formula of the Lorentz transformation.

(50) $x' = \frac{(c-v)t}{\sqrt{1-\frac{v^2}{c^2}}}$ (Einstein, 2009)

(51) $t' = \frac{\left(1-\frac{v}{c}\right)t}{\sqrt{1-\frac{v^2}{c^2}}}$ (Einstein, 2009)

Dividing (50) by (51) results in the system of equations (52) (Einstein, 2009).

(52) $x' = ct'$ (Einstein, 2009)

According to the formula (52), light moves at the speed c, which is the same with respect to the reference system K', regardless of the direction (Einstein, 2009). This corresponds to the formulas of the Lorentz transformation, as these were derived according to the constant speed of light c on the positive x-axis given in (49) and (52) (Einstein, 2009).

32 How moving rods and clocks behave

If a meter is defined on the X' axis between the two points x' = 0 and x' = 1 with respect to the reference system K', the question arises as to how long this rod is relative to the reference system K (Einstein, 2009)? Albert Einstein (2009) obtained the answer by considering where the beginning and the end of this meter rod are located relative to K at a fixed time t with respect to the same reference system K (Figure 18) (Einstein, 2009). Albert Einstein (2009) obtained the two points for the beginning and end of the rod from the formula (45) of the Lorentz transformation as soon as he set the time t equal to 0.

(53) $x_{(Start\ of\ rod)} = 0 \times \sqrt{1 - \frac{v^2}{c^2}}$ (Einstein, 2009)

(54) $x_{(End\ of\ rod)} = 1 \times \sqrt{1 - \frac{v^2}{c^2}}$ (Einstein, 2009)

The distance between these points is $\sqrt{1 - \frac{v^2}{c^2}}$ (Einstein, 2009). This meter rod moves relative to the reference system K not with c but with velocity v (Einstein, 2009). The rigid rod, which moves longitudinally with velocity v, has the length $\sqrt{1 - \frac{v^2}{c^2}}$ meters (Einstein, 2009). An immovable rigid rod is therefore longer than the same rod when it is in a state of motion, which means that the rod becomes shorter and shorter as the velocity v increases (Einstein, 2009). If the velocity v of the rigid meter rod is equal to c, then $\sqrt{1 - \frac{v^2}{c^2}}$ would be zero, making this root imaginary for velocities higher than c (Einstein, 2009). This is the reason why, in the special (and general) theory of relativity, the speed of light c represents a maximum possible speed, a so-called limit speed, which cannot be achieved by any massive body and therefore cannot be exceeded (Einstein, 2009).

The limiting speed of light c is a result of the formulas of the Lorentz transformation, as these become meaningless as soon as v > c is defined (Einstein, 2009).

According to the principle of relativity, in the opposite case for a rigid meter rod that is immobile relative to the reference system K on the X-axis, its length would also be $\sqrt{1-\frac{v^2}{c^2}}$ meters when viewed from the reference system K' on the X'-axis (Einstein, 2009).

The values for x, y, z, and t correspond to measurement results that can be obtained for comparative scales and simultaneity measuring devices such as clocks (Figure 18), therefore their physical behavior does not have to be experienced ex post from the Lorentz formulas (Einstein, 2009).

As an example to illustrate the physical behavior of rods and clocks, Albert Einstein (2009) defined a clock based only on seconds as a simultaneity meter (Figure 18), whose hand jumps once to one and then jumps back to zero. The jumps of this clock hand, which always rests at the starting point x' equal to zero with respect to the reference system K', take place continuously alternating between t' equal to 0 and t' equal to 1 (Einstein, 2009). Formulas (45) and (48) of the Lorentz transformation can be used to determine (55) and (56) for the two clock hand jumps (Einstein, 2009).

(55) $t = 0$ (Einstein, 2009)

(56) $t = \frac{1}{\sqrt{1-\frac{v^2}{c^2}}}$ (Einstein, 2009)

This hand clock, or measured time, has the speed of movement v with respect to the reference system K (Figure 18), therefore slightly more than one second elapses between the two jumps, i.e. a slightly longer unit of time in the form of $\frac{1}{\sqrt{1-\frac{v^2}{c^2}}}$ seconds (Einstein, 2009). The same clock hand previously moved faster in the state of immobility and now this clock or time advances more slowly due to the movement

(Einstein, 2009). The limiting speed of light c is also not possible here and therefore represents a maximum speed that can never be reached, even for time (Einstein, 2009).

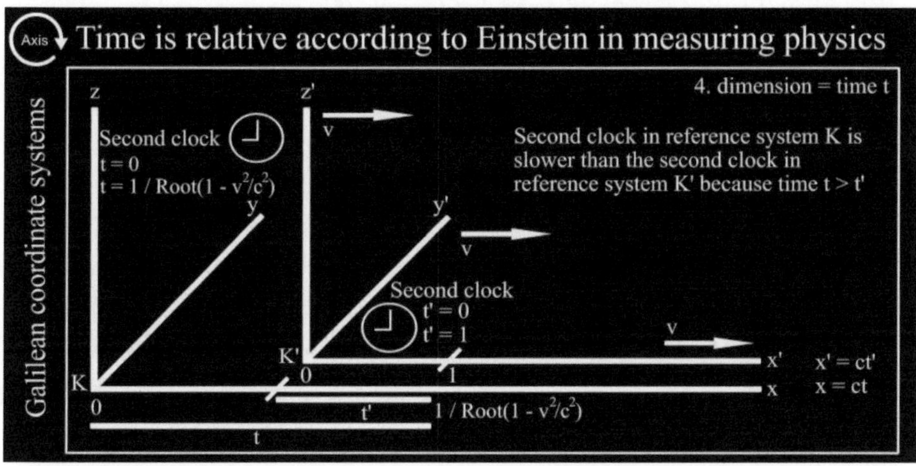

Figure 18. Time is relative according to Einstein in measuring physics (Einstein, 2009, p. 22).

33 Fizeau's experiment on the addition law of velocities

Due to the fact that in space-time clocks and rigid rods can only be set in motion at speeds lower than the speed of light c (Figure 19), the conclusions in section 32 will be extremely difficult to compare with reality (Einstein, 2009). Based on these conclusions, this section will derive another theoretical effect that has not yet been described and that can be verified, i.e. checked, experimentally in the best possible way (Einstein, 2009).

According to the assumptions of Newtonian mechanics, the law of addition of velocities was derived in section 26 for parallel movements of bodies (Einstein, 2009). According to the theory of relativity, Albert Einstein (2009) now assumed that a point (event) moves according to the formula (57) with the velocity w relative to the reference system K' (Figure 19).

(57) $x' = wt'$ (Einstein, 2009)

W corresponds to the velocity of the reference system K relative to K', whereas K' moves relative to K with v and the point (light body) relative to its reference system K' with w (Einstein, 2009). If x' and t' of formula (57) are expressed as x and t via the formulas (45) and (48) of the Lorentz transformation, then Albert Einstein (2009) obtained the equation (58) as a result, which corresponds to the law of addition of parallel velocities according to the theory of relativity.

(58) $W = \frac{v + w}{1 + \frac{vw}{c^2}}$ (Einstein, 2009)

The addition law of parallel velocities (58) corresponds to the theorem of experience, since (58) can be confirmed by Fizeau's experiment, which used the movement of light in a resting liquid as the test object and has already been repeated by other physicists (Einstein, 2009). In this experiment, the question is (Einstein, 2009): How fast does light move with a velocity w in a tube K (Figure 19) in the direction of the arrows when the tube K is flowed through by a resting liquid K' with velocity v?

Regardless of whether the liquid K' is moving relative to the tube K and the light or not (Figure 19), the movement of light relative to the liquid must always take place at the speed w (Einstein, 2009). The speed of light relative to the tube K must be determined because the speed of light relative to the liquid and the speed of the liquid relative to the tube K are given (Einstein, 2009).

The velocity W describes the movement of light relative to the tube K, so that formula (58) is valid and the Lorentz transformation is equivalent to reality (Figure 19) (Einstein, 2009).

The formula (58) established using the theory of relativity is confirmed to within one percent by the experiment and this reflects the effect of the fluid velocity v with regard to the movement of light (Einstein, 2009). This effect was also confirmed by electrodynamic experiments to evaluate hypotheses regarding the electromagnetic composition of matter by Lorentz (Einstein, 2009). This supports the evidence for the experiment described above, as Maxwell-Lorentz's electrodynamics do not contradict the theory of relativity (Einstein, 2009). The theory of relativity is therefore to be understood as a consequence of electrodynamics and represents a comprehensible résumé as well as a generalization of previously non-interdependent assumptions (Einstein, 2009).

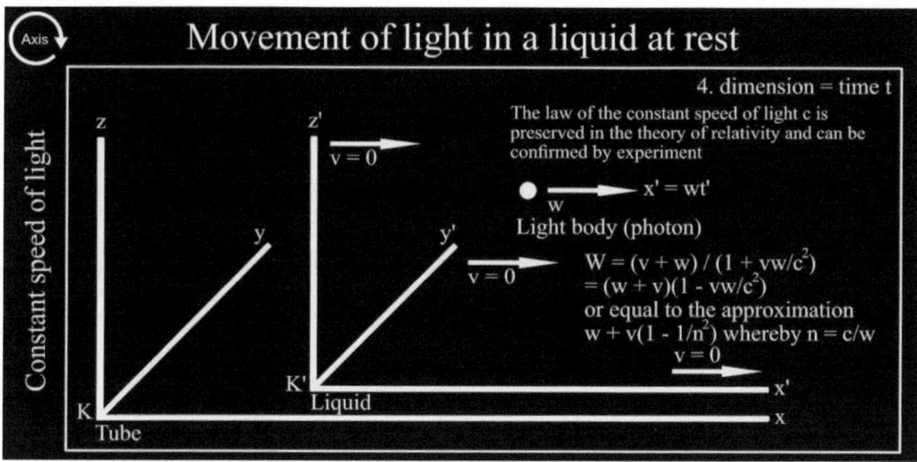

Figure 19. Movement of light in a liquid at rest (Einstein, 2009, p. 22 and p. 27).

34 The heuristic statements of the special relativity theory

As a summary of the previous sections 21 to 33, it can be stated that the principle of relativity as well as the speed of light apply in a vacuum and that the movement of light should be specified with the constant c (Einstein, 2009). If the relativity of the movement of light is considered, Albert Einstein (2009) obtained the Lorentz transformation theorem for the coordinates x, y and z as well as time t of the points (events), which together describe the natural phenomena, not as assumed in Newtonian mechanics (Einstein, 2009).

The law of the constant speed of light c is suitable for the description of natural phenomena, as it corresponds to the current state of knowledge (Einstein, 2009). With the Lorentz transformation, the law of the constant speed of light c can be combined with the principle of relativity to form a theory of relativity that can be summarized as follows on the basis of its heuristic statements (Einstein, 2009):

With regard to the Lorentz transformation, universally valid laws of nature exist covariantly (Einstein, 2009). All laws of nature are thus created in the same way and should therefore apply equally everywhere in space-time (Einstein, 2009). For space-time, this constancy of general laws of nature is expressed mathematically using the Lorentz transformation via the variables x, y, z and t of the initial reference system K (coordinate system) for the changed variables x', y', z' and t' of another reference system K' (Einstein, 2009).

The theory of relativity is a heuristic method for determining general laws of nature that should be valid in space-time, since the theory of relativity provides existing laws of nature with a fixed mathematical precondition (Einstein, 2009). In the event that a general law of nature were to be discovered that does not comply with this derived precondition, then at least either the principle of relativity or the law of the constant speed of light c, both contained in the theory of relativity, would be invalid and thus the theory of relativity would have to be rebuilt in a fundamentally different way (Einstein, 2009).

A modified fundamental structure of the theory of relativity, which contains this new general law of nature as a physically meaningful discovery, is the general-special theory of relativity derived by Erik Kolek (Section 38 below).

35 General results of the special relativity theory

The special theory of relativity is based on electrodynamics and optics (Einstein, 2009). Albert Einstein (2009) made few modifications to both theories, but their derivation was simplified and the number of hypotheses reduced. The Maxwell-Lorentz theory of electrodynamics is given greater evidence by the theory of relativity, even if the experiment would have confirmed a result worse than one percent (Einstein, 2009).

The special theory of relativity required a change in Newtonian mechanics for its derivation (Einstein, 2009). In Newtonian mechanics, only the laws concerning fast body movements were changed, in which the speed v of physical bodies comes close to the movement of light (Einstein, 2009). According to experience, such fast movements only exist for quantum bodies, i.e. very small bodies such as ions and electrons (Einstein, 2009). The general theory of relativity becomes relevant for explaining the motion of macro-bodies, such as stars, which means that the special theory of relativity loses its significance with regard to the motion of very large bodies (Einstein, 2009). According to the special theory of relativity, the kinetic energy of a body with mass m is described by the formula (59) (Einstein, 2009).

(59) $mc^2(\frac{1}{\sqrt{1-\frac{v^2}{c^2}}} - 1)$ (Einstein, 2009)

As soon as the speed v of physical bodies becomes more and more similar to the speed of light c, formula (59) becomes infinite (Einstein, 2009). It can therefore be assumed that the velocity v of physical bodies should always remain less than the speed of light c, regardless of how much energy is available for the acceleration (Einstein, 2009). The boundary acceleration problem can also be seen when a series (60) for the kinetic energy is formed from (59) (Einstein, 2009).

(60) $mc^2 + m\frac{v^2}{2} + \frac{3}{8}m\frac{v^4}{c^2} + \cdots$ (Einstein, 2009)

If $\frac{v^2}{c^2}$ approaches 1 in (60), then the third expression from the theory of relativity is always less than the second expression, which is only relevant in Newtonian mechanics (Einstein, 2009). In mc^2 the velocity is not taken into account, which means that the first expression (60) is not suitable for describing the energy of a massive body as a function of velocity (Einstein, 2009).

The clarification of the definition of mass represents the most groundbreaking general result of the special theory of relativity (Einstein, 2009). Before the theory of relativity, physics knew two fundamentally important and independent conservation laws: The law of conservation of energy and the law of conservation of mass (Einstein, 2009). These conservation laws were combined into a single law of energy and mass in the special theory of relativity (Einstein, 2009).

The principle of relativity requires that the law of conservation of energy must continue to apply to all Galilean coordinate systems (K, K', K" etc.) as long as there is uniform motion, i.e. translation, between the systems (Einstein, 2009). With regard to the transfer between these reference systems, the Lorentz transformation is decisive (Einstein, 2009).

Based on these assumptions and with the help of Maxwell's electrodynamics, Albert Einstein (2009) came to the following conclusion: If a moving body with a (constant) velocity v absorbs the energy E_0 through absorbed radiation without a change in velocity, then the body-related energy increases with the value (61).

E_0 is the energy gained by a moving body, which is assessed on the basis of the coordinate system, which is also moving uniformly (Einstein, 2009).

(61) $\frac{E_0}{\sqrt{1-\frac{v^2}{c^2}}}$ (Einstein, 2009)

The energy of physical bodies (62) can therefore be determined by the kinetic energy (59) (Einstein, 2009).

(62) $\frac{(m+\frac{E_0}{c^2})c^2}{\sqrt{1-\frac{v^2}{c^2}}}$ (Einstein, 2009)

A body therefore has the same energy E as a flying body with speed v and mass m + E_0/c^2 (Einstein, 2009). If the body receives an energy E_0, the associated inertial mass m increases by E_0/c^2, i.e. the inertial body mass m is not constant and is therefore variably dependent on the individual energy adjustment (Einstein, 2009). An inertial body mass m as a reference system K is a quantity with respect to the associated energy E (Einstein, 2009). The law of conservation of the mass of a reference system corresponds to the law of conservation of its energy as long as the reference system does not absorb and release energy (Einstein, 2009). If formula (62) is converted to (63) to describe the energy, mc^2 describes the energy E of the body (still) without the absorption of E_0 (Einstein, 2009).

The body must be viewed from the reference system with the same motion (Einstein, 2009).

(63) $\frac{mc^2 + E_0}{\sqrt{1-\frac{v^2}{c^2}}}$ (Einstein, 2009)

The formula (63) temporarily eludes observability by experience, since the energy adjustments E_0 associated with the respective reference systems appear too small, so that adjustments to the inertia of a coordinate system are barely perceptible (Einstein, 2009). The mass m available before the energy adjustment is (now) too large compared to the expression E_0/c^2 (Einstein, 2009). For this reason, the law of conservation of mass with its own validity was successfully developed (Einstein, 2009).

A direct, present action at a distance according to Newton's gravitational theorems does not exist due to the breakthrough of Faraday-Maxwell's explanation of electrodynamic action at a distance by means of processes with a finite speed of motion in between (Einstein, 2009). In the theory of relativity, this instantaneous action at a distance with an infinite speed of motion v is always replaced by the action at a distance propagated at the speed of light c (Einstein, 2009). This exchange to a long-distance effect with a finite speed of motion c takes place due to the important significance of the speed of light c for the theory of relativity (Einstein, 2009). In the second section on the general theory of relativity, it was explained to what extent this action at a distance can be changed as a result (Einstein, 2009).

36 Evaluation of the special relativity theory through experience

Any experiential evidence, for example, the results of a fundamental experiment such as Fizeau's, solidifies the special theory of relativity, which also supports the original Maxwell-Lorentz hypothesis of electromagnetism (Einstein, 2009). Albert Einstein (2009) considered it extremely groundbreaking to write that the entire light spectrum of the fixed stars is influenced by the relative motion of the earth with respect to these fixed stars and that this corresponds perfectly with the experience gained according to the theory of relativity. Albert Einstein (2009) referred here to aberration, i.e. the movement of the earth in its orbit around the sun, which causes a gradual change in the supposed locations of fixed stars each year, and to the effect on the color of starlight due to the radial component of the relative orbital movements of these fixed stars with respect to the earth. The light color effect (Doppler effect) is shown by a slight translation (displacement movement) of the resonance lines (spectral lines) of this fixed star light compared to the spectral position of the same resonance line produced with an illumination located on Earth (Einstein, 2009). Furthermore, there are all kinds of results from experiments that refer to the electrodynamic theory according to Maxwell-Lorentz and therefore also have a validity for the theory of relativity, which do not confirm any other theory by means of experience (Einstein, 2009).

According to Albert Einstein (2009), two types of results have since been obtained from experiments that the electrodynamic theory according to Maxwell-Lorentz is only able to describe by means of an additional auxiliary hypothesis, which actually seems strange if the theory of relativity is not used for this purpose.

β-rays, which are emitted by substances through atomic decay, as well as cathode rays are both known to consist of negatively charged electrical bodies, the so-called electrons, which have an extremely low inertia and high velocity (Einstein, 2009). The principle of motion of these bodies (electrons) can be fully verified by analyzing the effect of electric and magnetic fields on the displacement of electron radiation (Einstein, 2009).

According to Albert Einstein (2009), electrodynamics did not contain any statement about the extent to which the movements of electrons can be described by a theory. If electric masses with the same sign push each other away, then electrons, meaning rather their determining negatively charged electric mass, should be pulled away from each other due to the effect of the associated interaction, unless there could be interactions of another class acting between the electric masses, which have since remained obscure by their nature (Einstein, 2009).

According to the general theory of relativity, electrons or their electric masses are prevented from moving apart by gravity as a fundamental interaction (Einstein, 2009).

A theory of electron motion that cannot be explained by experience arises when the assumption is made that the relative distances between the determining electric masses of an electron can continue to exist in the same way during its motion, i.e. a constant concatenation between the masses is assumed according to Newtonian mechanics (Einstein, 2009). Using completely conventional thoughts, Lorentz first succeeded in developing a consistent theory so that an electron body experiences a contraction in its direction of velocity due to its movement in relation to the formula $\sqrt{1 - v^2/c^2}$ (Einstein, 2009). As a theory that is consistent but not justified by electrodynamic effects, this corresponds to the principle of electron motion, which has been verified by experience with a high degree of accuracy (Einstein, 2009).

The special theory of relativity leads to the same principle of motion of electrons and dispenses with an additional auxiliary hypothesis about their structure and changes (Einstein, 2009). The same facts were evident in section 33, where Fizeau's experiment was described, which confirmed the theory of relativity through experience, without the need to develop assumptions about the physical nature of the fluid at rest (Einstein, 2009).

The question of whether the movement of the Earth in space has an influence on an experiment on Earth describes the second type of physical results from experiments (Einstein, 2009). It has already been shown in section 25 that every experiment of this kind had a negative result (Einstein, 2009). Before the theory of relativity was developed, it was difficult for science (in general, not just physics) to be confronted with such a negative result because the facts were as follows (Einstein, 2009): The knowledge gained about space-time has left no possibility open, so that for the change from one reference frame K to the next reference frame K' the Galilean transformation is decisive (Einstein, 2009). If Albert Einstein (2009) assumed that the Maxwell-Lorentz formulas are valid for the reference system K, then he learned that they are invalid for a uniformly moving reference system K' relative to K if he assumed that the expressions of the Galilean transformation exist between the coordinate points x, y, z, t and x', y', z', t'. This could give the impression that one of each of the Galilean reference frames, in this case K, could be physically described by a fixed state of motion (Einstein, 2009). This result was physically interpreted by scientists to mean that K was considered relatively immobile to a theoretically existing aether light (Einstein, 2009). In comparison, any reference system K' would have to move relative to K and appear to move against this aether of light (Einstein, 2009). The K' movement against the light aether (K' relative to the "aether displacement") was ascribed more difficult natural laws, which would have to be valid relative to K' (Einstein, 2009). Consequently, a corresponding aether displacement had to be assumed for the earth's motion as well, because this was the aim of the scientists' efforts to prove the aether for a while (Einstein, 2009).

Michelson had discovered a possibility for the proof of aether that could not fail (Einstein, 2009). Mirrors A and B are attached to a rigid body with their reflective coatings facing each other (Einstein, 2009). If this entire reference system is immobile with respect to the aether light, a light signal needs a fixed time T to travel from mirror A to mirror B and from mirror B back to mirror A (Einstein, 2009). If this body including mirrors is moving relative to the light aether, Albert Einstein (2009) found (by means of calculation) a minimally different time T' for this outward and return journey of the light. Indeed, this calculation also yields ample results (Einstein, 2009). For a given velocity v of the body against the light aether, the time T' would be a different velocity w if this body moved vertically to its mirror surfaces A and B than if it moved horizontally to its mirror surfaces A and B (Einstein, 2009). Even though the calculated deviation between the two time periods was minimal, Michelson and Morley nevertheless carried out an experiment to determine the superposition (interference), in which this deviation should have been precisely apparent (Einstein, 2009). To the great uncertainty of the scientists, however, the experiment yielded a negative result (Einstein, 2009). Lorentz together with Fiz Gerald removed this uncertainty from the theory by concluding that the movement of the body against the light aether caused a shortening (contraction) of the body in its direction of velocity, because this contraction of the body was intended to achieve the loss of the stated time deviation (Einstein, 2009). The comparison with the descriptions of section 32 illustrates that this conclusion was also flawless according to the theory of relativity (Einstein, 2009). However, this physical understanding is a uniquely more satisfactory understanding according to relativity (Einstein, 2009). According to the theory of relativity, there is no reference system to be favored that fixes a ground for the implementation of the aether idea, therefore there is no aether displacement as well as no attempt that can bring the aether into evaluation (Einstein, 2009). The basic magnetic and electrical components of the theory also explain the contraction of moving bodies in the theory of relativity without requiring any further special assumption (Einstein, 2009). According to Albert Einstein (2009), body contraction was not dependent on motion, as this alone cannot explain contraction, but on the motion of the body with respect to a reference system. For the body including mirrors from the experiment by Michelson and Morley, this physical fact means that if the reference body K' including mirrors is moving with the planet K, such as the Earth, then no shortening (contraction) takes place, but in the case of a reference body K_0

(fixed star) relatively stationary with respect to a star, such as our Sun (Einstein, 2009).

37 The four dimensions of Minkowski space

Usually, the non-physicist is gripped by a puzzling fear as soon as he or she hears about something like four dimensions, an emotion close to that evoked by a movie ghost (Einstein, 2009). Nevertheless, no principle is easier to understand than the one that describes our experienced environment in terms of four dimensions as a time-space event (continuum) (Einstein, 2009).

A space represents a continuum with three dimensions (Einstein, 2009). This means that people are able to specify the position of a (stationary) point with three coordinates (numbers) x, y, and z (Einstein, 2009). With regard to these points, there may be variably adjacent points whose position can be described by the coordinate numbers x_1, y_1 and z_1, which can variably approximate the values x, y, and z of the first point (Einstein, 2009). Due to this coordinate approximation (property), the term continuum is used because of the three numbers contained in the coordinate system – the three dimensions (Einstein, 2009).

Accordingly, an environment of events based on physics (the continuum), referred to by Minkowski simply as the universe, naturally appears with four dimensions in a time-space meaning (Einstein, 2009). This is due to the fact that causality and thus our entire universe is made up of separate moments (events), each of which can be specified with four coordinates, the three spatial numbers x, y, and z as well as a time number t (Einstein, 2009). A universe also represents the continuum, since each moment has variable adjacent (realized or still possible) moments with the coordinate numbers x_1, y_1, z_1, and t_1, which are infinitely close to the coordinates of an initially observed moment x, y, z, and t (Einstein, 2009).

People are not adapted to this new improved understanding of time to perceive their environment as an event (continuum) with four dimensions. The reason for this is that time as a phenomenon in physics before the development of the theory of relativity has a different, rather fixed, i.e. absolute, meaning with regard to spatial coordinates (Einstein, 2009). For this reason, people did not adapt to this dynamic understanding of time, but (continue to) perceive time as a continuum (event) that exists separately from space (Einstein, 2009). In fact, according to the theory of relativity in Newtonian mechanics, this dynamic time corresponds to an absolute time that is not dependent on a location and a state of motion of a reference body (Einstein, 2009). This is taken into account in the Galilean transformation with the formula t' = t (Einstein, 2009).

Due to the (general) theory of relativity, the observation of the universe (including our environment for humans) becomes obligatory in four dimensions, because time no longer has any independence here, which is similar to the fourth formula of the Lorentz transformation [(48) = (64)] (Einstein, 2009).

(64) $t' = \frac{t-\frac{v}{c^2}x}{\sqrt{1-\frac{v^2}{c^2}}}$ (Einstein, 2009)

According to the latter formula, the time deviation Δt' between two moments with respect to K' also does not resolve in general if the time deviation Δt of these two events with respect to K resolves (Einstein, 2009). A time difference between two moments with respect to K' arises as a consequence of a spatial distance between these two moments with respect to K alone (Einstein, 2009). This finding is also not the central result of Minkowski, which is necessary for the development of a theory of existential relativity in the universe (Einstein, 2009). The result is based on the fact that in the theory of relativity a continuum with four dimensions, including the associated properties to be observed, is closely related to a continuum with three dimensions as in Euclid's geometry (Einstein, 2009).

According to Albert Einstein (2009), in four-dimensional Minkowski space $x^2 + y^2 + z^2 + t^2 = x_1^2 + x_2^2 + x_3^2 + x_4^2 = x'_1{}^2 + x'_2{}^2 + x'_3{}^2 + x'_4{}^2 = x'^2 + y'^2 + z'^2 + t'^2$.

To make the relationship stronger, the usual time coordinate t should be exchanged for the unit $\sqrt{-1}ct$ (Einstein, 2009). This exchange of time values ensures that the requirements contained in the special theory of relativity are mathematically complied with by the laws of nature, as these assign the same meaning to the time coordinate t and the three spatial coordinates x, y and z (Einstein, 2009). Strictly speaking, all four coordinate values together correspond to the three spatial coordinates in Euclid's geometry (Einstein, 2009). This makes the (special) theory of relativity much easier to understand, which should also be apparent to anyone who is not a mathematics enthusiast (Einstein, 2009).

The brief explanations merely illustrate a rough idea of Minkowski's central content; if these were missing, the general theory of relativity described above would probably have remained under development (Einstein, 2009), as would the general-special theory of relativity described in the fourth section. Since Minkowski's difficult-to-understand expressions are not necessary for a better understanding of the basic assumptions of general and special (as well as general-special) relativity by a mathematically inexperienced reader, (not only) Albert Einstein (2009) (but also Erik Kolek) wanted to refrain from a deeper mathematical consideration of this subject (Minkowski space) at this point, so that it can be referred to again in the concluding descriptions of this book.

Fourth section: On the general-special relativity theory

38 General-special relativity principle

The special principle of relativity, which describes any uniform rectilinear motion of a body, applies in particular to the law of the constant speed of light c (Einstein, 2009). On the contrary, the general principle of relativity states: Light generally moves along an arc line within a gravitational field and generally heavy inertial masses are accelerated unevenly through this field (Einstein, 2009).

As readers will recall, either the relativity principle or the constant speed of light principle c should be true. In order to find this out according to experience, the general-special relativity principle (by Erik Kolek) is derived below (Figure 20) by inferring the special relativity principle (both by Albert Einstein) from the general one. The new general-special relativity principle derived here should therefore continue to include the general and special relativity principles according to a supersymmetric and relativistic both-as-well-as logic.

Since light is generally not a heavy inertial mass, light is also accelerated uniformly on an arc line in a gravitational field at the speed c and this applies in accordance with the general and special principle of relativity according to Albert Einstein (2009). In addition, the constant speed of light c is equated in practice with a maximum speed of heavy inertial bodies; even now, a question of existence between relativity and the uniform speed of light c should raise justified doubts.

If, for example, I imagine a binary star system without planets, in which two stars K and K' of exactly the same nature rotate around each other, but move together in one direction at a speed w that is equal to the (approximately) uniform speed c, because the heavy mass of a star can certainly be considered equal to an inertial system K or K', it is immediately apparent or appears observable in accordance with experience how the speed of light c should behave in physical reality according to the general-special theory of relativity.

The inertial systems behave in the same way as K to K' analogous to K' to K'', i.e. K'' behaves in the same way as K. The movement of the bodies is represented by Galilean coordinate systems, which move in a Gauss coordinate system that represents the gravitational field. Since every light body (or every photon) must have the speed c regardless of its direction according to the special theory of relativity, which integrates the law of the constant speed of light c regardless of the direction of the light-generating radiation source, this also applies to the light of the respective star system emitted in the same direction according to the general theory of relativity. From the observation point of view of the light body in relation to its star, W = v + w always applies, i.e. W = c + c, then added together results in W = 2c, here the time t is always equal to 1.

This result should not be surprising according to Albert Einstein (2009), because according to Newtonian mechanics, the expression w = c – v or transformed to c = w + v indicates that c cannot be constant and can therefore only be variable, i.e. there should actually be no limit to the speed of light. Accordingly, if a light beam is emitted with the speed w or w' in the same direction as its emitting reference body, which moves with the speed v or v', then v = 0 can be assumed for the sake of simplicity, whereby w = c, but this is already a relative consideration of the speed of light c according to the definition of simultaneity. If, on the other hand, we think classically and choose v' = 1, then 1 + w' = c', whereby c' > c and therefore a constant speed of light c is unlikely even according to Newton's theory of motion.

The law of addition of velocities in relativistic mechanics states that $W = (v + w) / (1 + vw/c^2) = (c + c) / (1 + cc/c^2) = 2c / (1 + c^2/c^2) = 2c / 2 = c$. However, this c merely means that the two Galilean coordinate systems under consideration move constantly at the speed c within the Gauss coordinate system, i.e. in the gravitational field on an arc line. However, light in this general-special case, considered individually, according to the general-special theory of relativity, moves at 2c; thus the law of the constancy of the speed of light c is no longer tenable, according to experience, and must be modified or referred to as the law of the simultaneous constancy of the relative speed of light c, which can be different for individual bodies, but appears constant when considered simultaneously according to the general-special theory of relativity, which observes the general and special principle of relativity.

It should now be clear that there is no limitation of speed in the universe, not even in the case of light, therefore c = v = w. This results in the new adjusted energy expression $(mv^2 + E_0) / \sqrt{(1 - w^2/v^2)}$. In even more familiar notation without initial energy E_0, this can be expressed as: $E = mv^2$. The previous statement $E = mc^2$ thus appears to be improved in the general-special theory of relativity, which is physically possible because the law of the uniform speed of light c was merely a basis for Albert Einstein (2009) to derive his special theory of relativity. This uniform rectilinear motion was then made non-uniformly curvilinear by Albert Einstein (2009) in his general theory of relativity with the help of a gravitational field system in accordance with the spatial state of our universe.

The previous expression $E = mc^2$ is still valid in the general and special theory of relativity (Einstein, 2009), but in the general-special theory of relativity $E = mv^2$ applies. The limiting acceleration problem for the kinetic light energy according to Albert Einstein (2009) can thus be adapted as follows: $mv^2 + m\frac{w^2}{2} + \frac{3}{8}m\frac{w^4}{v^2} + \cdots$ or supplemented by the temperature T, because the heat of light is a component of the mass of light: $Tmv^2 + Tm\frac{w^2}{2} + \frac{3}{8}Tm\frac{w^4}{v^2} + \cdots$ (Einstein, 2009). If $\frac{v^2}{w^2}$ approaches 1, then the second expression from Newtonian mechanics is always greater than the third expression from the theory of relativity (Einstein, 2009). The velocity is taken into account in mv^2, which means that the first term is now suitable for describing the kinetic energy of a massive body as a function of velocity (and temperature) (Einstein, 2009).

However, Fizeau's experiment (Section 33) proves the speed of light c. This appears to be correct only in so far as this is a single experiment on Earth and not an experiment carried out simultaneously on two celestial bodies, so the speed of light could be different on Mars, for example, compared with that on Earth, as these are two different gravitational fields. Now, however, a measured difference in the speed of light should be analogous to the difference in the gravitational field or be present accordingly; this makes it seem possible to determine the actual gravitational field strength even more precisely in physical mathematical terms.

In addition, however, it also seems conceivable that the human eyes could not see any acceleration of light or only approximately up to the speed c of light, which means that all movements of light, which then appear faster to us, could therefore be seen by us as black; all available measurement results for the constant speed of light c would then be mentally influenced by this, because according to optics no truth could be seen by us, in particular because of this a false sensory impression about the constant speed of light c could be possible. This probably unnoticed but conceivable sensory limitation of the human eyes could have influenced an experience gained from the experiment according to Fizeau, as a result of which not the constant speed of light c but our maximum visual speed c would then have been falsely determined; this is conceivable in principle because optics is part of electrodynamics and thus part of the (general-special) theory of relativity.

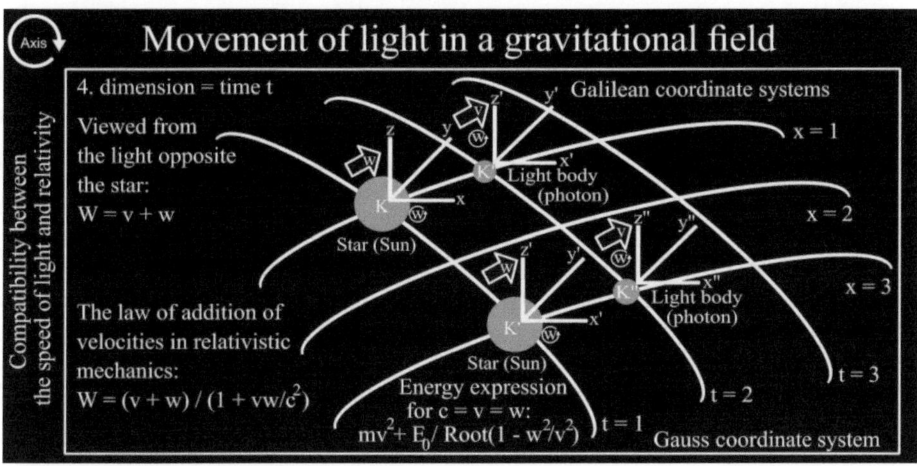

Figure 20. Derivation of the general-special relativity principle.

39 The field of light propagation

If I imagine a universe in which a single large gravitational field exists due to countless stars, then it is clear that there could be an interaction between light and darkness. If I mentally equate the gravitational field with a light propagation field, then it becomes clear, if we generally imagine that nothing else would exist in the universe, that light, viewed as a multidimensional field, should have a different luminosity (energy intensity) at different event points in the continuum. Due to the physical fact that stars exist everywhere in the universe, light also exists localized accordingly – independent of human perception as a color without sound in a vacuum – it can therefore always have an influence on space combined with time and all bodies contained therein.

If light is emitted from a star, it could be possible that, as shown by the law of the simultaneous constancy of the relative speed of light c, the speed v first represents an increasing acceleration and then a decreasing acceleration due to gravitational fields is conceivable. In this case, for example, accumulations of light emissions would be noticeable at the edge of our solar system, such as bright multi-colored auroras, which, however, should represent bright multi-red plasma phenomena that could possibly also be heard acoustically if these wave-like rippled gravitational field surfaces were to hit the membrane of a microphone. Since there is a red change in the spectral straight line due to the potential effect of the gravitational field of all bodies present in the star system, the general-special theory of relativity is consistent (Einstein, 2009). At the same time, the understanding of the change in the straight line, because this knowledge is now understood as a consequence originally based on the action potential of the gravitational fields, makes it possible to draw important conclusions regarding the masses of the bodies within our space-time continuum (Einstein, 2009), because according to the general laws of nature, darker shining stars should therefore be heavier (or more inert) than brighter shining stars.

The temperature of a light propagation field, starting from the star to the furthest possible radiation distance point, should decrease non-Euclidean in geometry in the respective four-dimensional space-time regions up to the temperature of the acoustic background noise, which the researchers Penzias and Wilson discovered as early as 1964 and associated with a possible Big Bang of our universe. On the other hand, this background noise could also have always been a phenomenon in our universe that has always existed in connection with stars and light. However, the strength of the background noise could vary. This could depend on the respective space-time region. A star, i.e. its simultaneous nuclear fission and nuclear fusion processes of chemical elements, should not be audible in a vacuum, not even a reaction like a super nova could be audible there. This only means that in this analogy a background noise, which should be related to the Big Bang, as an analog for the formation of our universe in general, should also be due to a differently tempered light propagation field (general-specific gravitational field).

What could a gravitational field consist of or how could people also perceive this energy-related phenomenon? Physicists have known since Newton that an apple and all other objectified bodies can experience gravitational attraction through this interaction. Not only physicists should consider what other things could experience an attraction and how this attraction could influence our world according to geometry? So how should we imagine this world in physical reality?

As we have learned so far, Albert Einstein (2009) first assumed a spherical geometry, then an elliptical geometry and then even a manifold geometry for our space-time continuum. If each of these spatial states for our universe were to resemble a spatial state in our natural environment, then all objectified bodies would be spherical, elliptical or even manifold. A geometrically square shape, i.e. as people imagine it, such as a cube or cuboid, and otherwise deviating shapes (e.g. pyramid) could therefore still have a spherical, elliptical or even manifold proportion in individual areas as soon as a closer look is taken. To express this more figuratively, for example, a book like this one could be geometrically curvilinear, partly non-Euclidean and also partly Euclidean. Euclidean and non-Euclidean areas could alternate optically, and corresponding angular shifts would then have to be taken into account mathematically if necessary.

In the general-special theory of relativity, no uniform straight lines are (also) assumed to exist in nature. However, this does not mean that no uniform straight lines could exist in physical reality; these would have to be created artificially using Euclid's geometry; by mathematically adjusting the angle in the relevant area. The spirit level could not be used as a measuring instrument for this, because it could provide inaccurate measurement results when applied to spheres or ellipses, because although an actual geodesic line could be mentally interpreted as a straight line, it cannot show the difference in angle, i.e. the distance to the spherical surface (or ellipsoidal surface). This would actually result in architectures whose stability would be unsatisfactory due to a lack of spherical geometry, elliptical geometry or manifold geometry. A new kind of spirit level, which makes the differences in two or more points visible for the imaginary uniform straight line, would be, so to speak, the right measuring instrument for representing the spherical metrics on a celestial body such as the Earth.

Such a light propagation field should always be present, it could therefore have an energy-transmitting effect on objectified bodies. The (kinetic) energy could be equal to the gravitational field and could be due to the space-time state of our universe. So where there is a gravitational field, there could be (kinetic) energy, which could also appear to be usable, for example by a gravitational field power plant.

40 The equality of rigid and non-rigid reference bodies as a justification of the general-special relativity principle

In the general-special theory of relativity, reference bodies are assumed to be both practically rigid and non-rigid bodies. In this theory, this is due to its components: special relativity mechanics assumes rigid bodies and general relativity mechanics assumes non-rigid bodies. So to what extent can there be an equality between these two reference bodies and their motion?

For our universe, we have to imagine that, according to Newton's theory, all bodies are in motion, i.e. they are not rigid and, in contrast, the other extreme is also conceivable, namely that no objectified body could be in motion, which I call "modern mechanics". The equality of these two types of reference bodies (rigid vs. moving) could come about because, on the one hand, a movement is a movement, as people can see it, and therefore also be something else at the same time, such as a reflection of a single space-time state that can only be seen briefly. At the same time, however, this would also point to a much higher dimensioning of our universe. Objectified bodies could thus exist simultaneously in several dimensions without their physical reality having to be influenced by this.

Although there is no absolute time, the individual space-time states could therefore only exist fleetingly for a single moment. The individual moments could be perceptible, analogous to reflections that people could not have consciously perceived up to now, this is based on "modern mechanics". If I imagine the movement of a mirrored body of light back and forth at up to infinite speed of light v, i.e. the beam of light constantly oscillates back and forth between a starting point and an end point, then according to Newtonian mechanics I should only see a bright spot at the beginning and end of the movement. Or, according to modern mechanics, I should see two bright event points between the beginning and the end as well as an infinite number of coordinates in between that only appear very briefly and are therefore less brightly visible. If, for example, I move my forearm back and forth as fast as I can in this physical sense, then according to this experience I cannot make a clear assignment to classical mechanics according to optics, but also not to modern mechanics, which is why both space-time states remain conceivable or remain simultaneously.

So if a body (like a planet) moves around a star (like our sun) within a multi-dimensional elliptical orbit (as in general relativity with 3 space coordinates + 1 time coordinate or as in general-special relativity with 3 space coordinates + 3 time coordinates + one coordinate dimension for the curvature intensity) starting from an arbitrarily determined point event, which simultaneously marks the start and end of the movement, then in the case of absolute time (Newtonian theory) it could be that this body does not move at all in physical reality (x, y, z) due to the universal time (t) of the universe. On the other hand, in the case of non-absolute time (Einstein's theory), it could also be quite possible that this body does move in physical reality (x, y, z), but only due to the time (t) firmly connected to space at an infinite number of extremely short but simultaneously appearing coordinate points (x, y, z, t). A non-absolute time (Einstein's theory) could therefore mentally encompass an absolute time (Newton's theory) and thus not exclude it.

Readers should have noticed by now that this is a different concept of infinity, not only thought of in terms of spatial dimensions with a time dimension (Newton's theory), but instead the expansion is transferred to the spatial and temporal dimensions of our universe (continuum) (Einstein's theory). Therefore, every finite point in the infinite point neighborhood of our universe (continuum) could really exist finitely and infinitely at the same time. At this point, the mental complexity (difficulty) should be reduced again as soon as an abrupt point expansion (Big Bang of the universe) is imagined. Could this point explosion exist, as previously assumed, due to successive expanding space-distances at certain time intervals (Newton's theory) as well as simultaneously in a single moment between the initial event and consisting of a suddenly expanded space-time distance (Einstein's theory)? From a physical point of view, the answer to this question could indeed be yes, but only if no absolute time and no absolute space existed in our universe (continuum), because then the universe (continuum) could represent a space-time state reflection that could also have just appeared today (Einstein's theory). However, this single moment could be infinitely superimposed on the expanding space distances at certain time intervals (Newtonian theory).

41 To what extent are there principles in the general and special relativity theory that require a general-special relativity theory?

The foundations of the general and special theory of relativity have only minimal limitations, and this is due to the excellent derivation by Albert Einstein. The special theory of relativity mentally adopts the assumption about the constancy of the speed of light c (which contradicts relativity and is therefore actually wrong) only in order to explain uniform rectilinear movements. In the general theory of relativity, this movement becomes non-uniformly curvilinear, whereby a concentration on the gravitational field is mentally placed, but this makes an indirect realization of the law of the constant speed of light possible; this cannot be uniform and non-rectilinear on this basis alone, at least Albert Einstein (2009) gives enough evidence to question this general law of nature.

The present derivation of the general-special theory of relativity indicates that Albert Einstein (2009) already knew from the beginning that the constancy of the speed of light c is not mentally tenable, but it could have been difficult for him to directly express a different opinion due to today's generation of physicists, since they are convinced of their ideas about the physical world and can only be convinced of opposites step by step from theory to theory by means of geometric mathematical proof. A further indication of this possible idea is that Albert Einstein (2009) generally always preferred v instead of c to indicate the speed, unless this quantity c was already contained in equation systems such as the Lorentz transformation.

Most of the limitations regarding the general and special theory of relativity are not due to the work of Albert Einstein (2009), but to an inadequate and therefore incorrect understanding of his readers, especially from the field of physics, and because people in general seem to lack geometric mathematical imagination (form of perception). In particular, I have noticed this when, for example, a flat space-time area (like a stretched cloth) on which the bodies (therefore only two-dimensionally understood) would lie embedded is incorrectly spoken of or that clocks of the same nature would run faster (or slower) on moving bodies. Albert Einstein (2009) was only concerned with the movement of the clock hand in the metric sense of a time measurement by comparing the distances between the time units measured in the inner and outer parts of a moving rotating body, which therefore proportionally eccentrically generates a gravitational field and none that constantly follows from the center of the classical theory. The origin of these misunderstandings appears to be based on the inductive objective way of thinking combined with an either-or logic, which is primarily practiced in science, instead of generally accepting and upholding the deductive subjective way of thinking combined with a both-and logic, as Albert Einstein (2009) tangibly exemplifies.

Some conceivable reasons for such incomplete, i.e. incorrectly formed, understandings of the readers of Albert Einstein (2009) are similar to the examples and designations he uses for his descriptions of physical reality. This involves, for example, a platform and a train as well as an observer who is either waiting on the platform or is on the train; once a raven is even used symbolically to describe the movement. Unfortunately, I can't exactly imagine why all readers could not see a star and the light directly in this example, but they would have had to, because this is merely a system of equations (platform = star), which should make it easier to understand. Albert Einstein's (2009) mental requirement to establish this mathematical analogy himself therefore appears to be generally difficult for his readers, which means that for all his readers, a train station remains a train station and does not become a star radiating light, as is actually intended by the charming anecdote, from a train station radiating trains; too much visual imagination is required for this; at least as long as no textual reference is given accordingly. I noticed this on the basis of theory, because the movement is described in relation to a quasi stationary and moving body and is useful for an imaginative representation of the constancy of the speed of light c.

A time dilation is generally assumed by the readers, but this is not described anywhere in the works of Albert Einstein (2009), so this incorrect understanding of time can only have arisen through incorrect interpretations of the readers with regard to the example of the train station by Albert Einstein; such a time dilation does not exist. The same misunderstanding on the part of readers of Albert Einstein (2009) could have arisen from the example with the clocks at rest in different places; here people could, due to their habits, generally think of time and its processes instead of a mechanically conceived time measurement as a clock, whose technical inner workings are influenced by the gravitational field just like the movements of the clock hands. This clock of the same nature, i.e. which has the same technical structure, would, for example, tell the time differently on Mars than on Earth according to measuring physics. The resulting relative difference in time measurement (time metrics) could have caused Albert Einstein (2009) to no longer believe in absolute time as in Newton's theory; if there is no absolute time, then time cannot be measured individually at points within our universe (continuum), regardless of whether clocks of the same nature are fixed (at rest) for time measurement there or not. Clocks can therefore only be used for simultaneous point events, but then the gravitational differences in the two space-time areas (event moments) within the reference systems (systems of equations) must be taken into account.

If I look multidimensionally at these real physical images in my head, which arise when I read works by Albert Einstein (2009), then these are always different from the now imagined images of another reader of Albert Einstein (2009), which should be justified by the purely symbolic, i.e. textual and mathematical representation of Albert Einstein (2009) in a context with only four illustrations. This is one of the reasons why I use supersymmetric, relativistic modelling and visualization to support the generation of truth (experience) for the didactic preparation of difficult geometric mathematical relationships. The viewers should get easier access to the generalized specialized theory and this should enable them to share generated knowledge (experience) more easily. As long as people can only communicate their real physical understanding to other people via texts, mathematics and model visualizations, etc., a certain discrepancy (difference) will always remain due to the Lorentz transformation; this should be less physically pronounced in the case of general-special relativity theory than in the case of general and special relativity theory.

42 Several conclusions from the general-special relativity principle

Some of the most far-reaching conclusions from the principle of general-special relativity are that the speed of light is not conceivable in physical reality and that light exists as a field everywhere in the universe (continuum).

If, for example, I imagine a test reference body K accelerated at the speed of light or even faster than light, which follows its geodetic course past a star (such as our sun) according to its navigational calculation, this body must take into account the gravitational field influence of this heavy inertial mass, since it must experience a deflection (at the latest) after the star (or the sun). This deflection caused by the gravitational field must be compensated for depending on which multi-dimensional space-time coordinates K' are to be achieved. For this purpose, at least 4 + n but a maximum of 7 dimensions (3 space coordinates, 3 time coordinates and an angle of curvature of the respective star system) are physically and naturally coincident and can be determined depending on the required precision of the results of the star navigation.

If this movement of the test reference body K still takes place according to Newton's theory, then it is clear that an absolute time defined as t = 0 and t = 1 must be assumed, so that a (practically) light-fast body with the speed v = c (t = 0) must radiate another light-fast body (for example a light particle or photon) in any selected direction of translation with constant speed w = c (t' = 1). Otherwise, the law of the constancy of the speed of light c would not be tenable (or otherwise, the law of the constancy of the speed of light c would be untenable (or conceivable) from the outset and, from the point of view of an observer resting firmly on the emitted body of light, the total speed 2c would result after $W = v_t + w_{t'}$, with which, according to experience, a new law on the simultaneous constancy of the relative speed of light c would have to be proven, which supplements and thus completes the old law on the constancy of the speed of light c on the basis of a modified truth; an imaginary speed limit v = c can therefore not exist in our universe (continuum).

If this movement of the test reference body K now takes place according to Einstein's theory, then it is also understandable that no absolute time defined as t = 0 and t = 1 is to be assumed, but since the metrically measured one-dimensional time is firmly connected with the three-dimensional space, x, y, z, t = 0, 0, 0, 1 for the initial event of the point emission W and x', y', z', t' = 1, 0, 0, 1 for the simultaneously occurring point emission relative to the initial event W'. These two four-vectors are necessary for the derivation of the simultaneous constancy of the relative speed of light c. For x, y, z, t = x', y', z', t' (and for seven dimensions x, y, z, t(x), t(y), t(z), α = x', y', z', t'(x'), t'(y'), t'(z'), α' or $x_1 + x_2 + x_3 + x_4 + x_5 + x_6 + x_7 = x'_1 + x'_2 + x'_3 + x'_4 + x'_5 + x'_6 + x'_7$), the modified Lorentz transformation is not used to determine the velocity c = w = v because this always results in zero for the coincident velocities v and w, in other words, it presupposes (or contains) a mathematically conceived speed limit in general for the movement of light and in particular for the movement of bodies. Instead, the system of equations $W = (v + w) / (1 + vw/c^2)$ is used for both instantaneous observations for the model representation of the two arbitrarily selectable moving space-time areas W = W'.

(65) $W = (v + w) / (1 + vw/c^2) = (c + 0) / (1 + c0/c^2) = c$ AND $W = v + w = c + 0 = c$

(66) $W' = (v + w) / (1 + vw/c^2) = (c + c) / (1 + c^2/c^2) = c$ AND $W' = v + w = c + c = 2c$

According to the general-special theory of relativity (taking Einstein's theory into account), the speed c = w = v always exists between two fast-moving event points W and W', whose speeds c and 2c, however, differ individually by c = w = v, whereby in general c is valid for two reference bodies W and W' and represents the law of the simultaneous constancy of the relative speed of light c, which extends the previous law with a new simple physical meaning and should therefore also be easy to understand. The mathematically conceived speed limit must be removed from the Lorentz transformation by modifying it into matching systems of equations. First, the Lorentz factor, given by $\Omega = 1 / \sqrt{[1 - (w/v)^2]} \geq 1$, must be used to determine a Lorentz factor that exists according to the general special theory of relativity. This is easy to express, since (according to the law of the simultaneous constancy of the relative speed of light c) a coincident speed v can exist for both bodies according to experience; in the general-special theory of relativity, therefore, only one valid space-time speed state is considered relative on the basis of two accelerated bodies. The relative Lorentz factor is therefore $\Omega = 1 / \sqrt{[1 - (1/v)^2]}$ for $0 \leq \Omega \leq 1$ (Kolek-Lorentz factor), here the velocity v is always to be regarded as a movement on the x-axis, which becomes positive by the specified square, even in the case of a backward movement, which can be determined analogously to a forward movement in a coordinate system. To specify individual steps in the speed of light, v = c can therefore be set and the time to represent the simultaneity of both velocity moments can be expressed with a positive sign; the initial event for this is v = 0 = c.

(67) $x' = \frac{x+t}{\sqrt{1-\frac{1}{c^2}}} = 1$ (Modified Lorentz transformation for the representation of space-time light velocity movements)

(68) y‘ = y = 0 (Einstein, 2009)

(69) z‘ = z = 0 (Einstein, 2009)

(70) $t' = \frac{t-\frac{1}{c^2}x}{\sqrt{1-\frac{1}{c^2}}} = 1$ (Modified Lorentz transformation for the representation of space-time light velocity movements)

If light is regarded as a field that exists everywhere in the universe (continuum), then it should have an influence on the masses in it (such as planets and the smallest particles) and (at least in part) be equivalent to the gravitational field of our universe (continuum). If light exists in the universe (continuum) as a field, it should be spread out differently in individual space-time areas, which means that darkness would then be an observable phenomenon that indicates that light does exist in this space-time area, but with a practically infinitely lower field energy intensity. Light emitted by stars should be able to push point masses into the space-time of our universe (continuum), thus providing a further influencing factor for determining the gravitational fields of bodies whose gravitational field makes itself felt from the inside outwards, for example by attracting free-flying bodies (such as asteroids) in a non-uniform curvilinear manner. This attraction is generally weaker the further apart the bodies concerned are. This light emitted by stars should also be able to have an influence on the direction of movement of bodies; freely flying bodies should thus receive a deflection after the light hits these bodies. Bodies rotating on elliptical orbits around a star (or even around several stars according to their common center of mass) could thus be attracted by the gravitational field of the heavy inert point masses, but at the same time be repelled due to the emitted light ray bodies. Such a field, which could consist of simultaneous attraction and repulsion of bodies in the space-time structure, can generally be described as an anti-gravitational field.

So if all bodies such as planets around a star (such as our sun) were to be exposed to an attractive but simultaneously repulsive gravitational light field, this could be calculated on the basis of the orbital motion of one or more planets around this star. The planet Mercury is ideal for this purpose, as it has already played an important role in confirming the general (and general-special) theory of relativity (Section 3.1). Astronomers can generally confirm the existence of the planet Mercury to this day. This means that Mercury can still be seen in the telescope during its solar transit and it does not appear to have been swallowed up by the sun. This leads to the question in this physical analogy: Why was the planet Mercury not attracted by the sun and could this still happen? To get to the bottom of this question, it can be expressed that although Mercury is closest (or most strongly) exposed to stellar gravity, there appears to be no non-uniform curvilinear attraction towards the sun, but equally there is no uniform rectilinear repulsion away from the sun (stellar antigravity) due to the radiation of the light body. This means that a balance between attraction and repulsion could exist at any point on Mercury's elliptical line around the sun or, in relation to the entire Mercury orbital curve, this balance could also only exist at one point as an absolute point event – at its perihelion point. Mercury, while moving on its orbit with the speed v, is therefore always exposed to the attractive gravitational field and the repulsive light field. Sometimes (probably every 100 years) there could be a fluctuation within this anti-gravitational field, which Albert Einstein (2009) also noted as a deviation from Newton's theory. If this gravitational field were stronger than the light field, it is conceivable that the speed v of Mercury would increase in a spiral around the sun until its final event, as would probably its temperature and mass. Observed perihelion shifts of planets close to stars like Mercury, understood as a general law of nature, could make a changed anti-gravitational field of stars visible, but this is only a theory about the relativity possibly generated by stars.

Relativity generally always refers to the reality of nature and the living beings in it, but in theory it is observed from several angles, such as a body moving away or approaching (like a car), the optical impression of which can always be assessed on the basis of an added circular diameter. How could anomalies of a gravitational field on a spherical surface be detected if they existed? Using the example of the movement of two electrons in the physically negative charge case, it can be said that they must move apart, but could be welded together by their gravitational field and yet the distances between them could be conditionally substitutable (variable). So if in nature the distances and thus the size appear relatively in each event point of a coordinate system, gravitational field anomalies could be determinable by arbitrarily selectable scales, because at these spherical surface points the distances between the electrons would have to be larger or smaller. However, depending on the intensity of the radiation, the spherical beings could experience that as soon as they are really in a possibly minimally (infinitesimally) brighter color-shifted field anomaly, the gravitational light effect itself could change their scale not only optically, but this would not be noticeable in their imagined physical reality (reality) without the relativity between light and gravitation theorized in this book; in this assumption about a dynamic anti-gravitational field, physically more (resp. less) light equals more (or less) gravity, but only in the physical case of radiation-absorbing reference bodies. According to relativity, the gravity of light-emitting bodies should become lower (or higher) if they emit more (or less) light. These assumptions about light (optics) as part of electrodynamics in the general-special theory of relativity could therefore establish an observable light field dynamics (optics dynamics) for these spherical beings, which could correspond to the invisible dynamics of the gravitational field of the space-time of their reference body in an interaction with the other (nearby) bodies in their (directly) perceptible world.

According to Albert Einstein's general theory of relativity (2009), Mercury's elliptical axis rotates like a straight line around the sun and this rotation is practically impossible to determine via the motion of the other planets in our system (Section 3.1), which is why we now only think of a gravitational light field with four dimensions in which Mercury and the sun are located. A practically existing equilibrium could be recognized at Mercury in a point that describes the rotational motion of the perihelion of the planet Mercury. The equation (71) by Albert Einstein (2009) can be understood as an equilibrium system between stellar gravity and stellar antigravity, which contains the quantity a, which represents the large semi-axis of the elliptical orbit. The e stands for the eccentricity (deviation from the circle), the c illustrates the speed of light in cm/second and T the orbital period in seconds (Einstein, 2009).

(71) $24\pi^3 \times a^2 / T^2c^2(1 - e^2)$ (Einstein, 2009)

To date, a perihelion motion has only been proven by astronomers for the planet Mercury. According to experience, this could mean two things: firstly, only the planet Mercury has this perihelion motion and no other planet (not even the Earth) has such a perihelion motion in physical reality. According to the general laws of nature, this is conceivable as a theory about the anti-gravitational field with four, six or seven dimensions, which means that the perihelion motion of Mercury demonstrates another important physical fact, namely a force-balancing anti-gravitational law according to the field concept of attraction g_{ik} = repulsion g_{ki} (four-vector). The field term for the more general light gravitational field described is therefore $g_{ik} - g_{ki} = 0$ due to the symmetry property, whereby the deflection of light after a star and the shift of the spectral line should be (even) more precisely calculable and explainable, which I do not want to go into in more detail mathematically in this presentation.

43 Behavior of scales and clocks on a stationary, non-rotating reference body

If I compare the behavior of scales and clocks by imagining two stationary non-rotating quasi-spherical reference bodies, regardless of whether such reference bodies really exist in our universe (continuum), on which these tools for measuring physics are stored, then I learn the following:

Although the arbitrarily selected scale itself is the same size on both bodies, it could still measure a different size distance in relation. Thus, a metric unit size could appear very long on a small body and very short on a large body, because this is due to the relativity of two bodies viewed simultaneously.

The same applies to a clock whose hands (not time in general) only appear to move faster on the small body and slower on a large body, at least if you look at the distances between the individual time periods on the clock face. On the small body, these intervals appear to be further apart on the clock face, which could make time seem to run slower. On the large body, these distances appear to be closer together on the clock face, which means that time could illusively run faster. However, a state of time only exists firmly connected to a state of space, as time is not absolute (Einstein's theory) and therefore it is not possible for any time to run faster or slower, as time is generally understood in this way: There is only one time. If we humans want to experience this one time, then it is always the infinitely finitely expandable dimension of our world; this also corresponds to Einstein's assumption from the fourth dimension onwards about the non-existent absolute time.

If there is such a thing as absolute time according to Newtonian mechanics, then in the general-special theory of relativity this would be the intrinsic time of space and all the bodies in it. A quasi-absolute time could therefore be understood as a single intrinsic time as well as a space-time equal to body-time, which would make a manifold concept of time conceivable, for example for sidereal time, planetary time and quantum time. If a clock of the same nature is firmly placed on both stationary non-rotating quasi-spherical reference bodies, then their different gravitational fields have an unequal influence on the clock hand due to a different amount of matter, which means that a large body slows down the clock more due to gravity, while a small body could make the clock run faster due to gravity. If a general expression is to be formulated for this, the following sentence could apply: *The more curved space is, the more curved is the time associated with space and vice versa*. Accordingly, an astrophysical macro-space-time (for example a stellar or planetary time) appears to be incomparable with a quantum-physical micro-space-time (for example a matter particle time) under these physical circumstances due to two gravitational field dimensions of different curvature; even in the physical case of the simultaneity of event moments, a temporal metric between a smallest and largest spatial state would (for the time being) be inconceivable.

44 A non-Euclidean and yet Euclidean universe (continuum)

If our universe (continuum) is viewed from the perspective of observers located on a planet, they should notice that space and everything in this space does not follow Euclidian geometry, i.e. it is not curvilinear, but on closer inspection it still appears Euclidean, i.e. uniformly rectilinear. So could a geodesic line, for example, contain straight lines as individual components? With regard to this question, we will remember that the derivation of Albert Einstein's theory of relativity (2009) starts with uniform straight lines (special theory of relativity), which are changed by a subsequent derivation to non-uniform curved lines (general theory of relativity). In the general-special theory of relativity, this was derived backwards via the non-uniform curved lines (general theory of relativity) to uniform straight lines (special theory of relativity), so in this theory both types of lines are simultaneously conceivable or objectified as body components. An arbitrarily selected reference body within the spatial structure of our universe (continuum) would have to be non-Euclidean as a whole and at the same time have Euclidean areas within its design if at least four to a maximum of seven dimensions are assumed for the universe, not just considered as a model.

To simplify things, you can imagine it like this: If we think of a banana as an example, the observers can imagine this banana as miniaturized as possible so that this macrobanana (macro-space-time) now becomes a quantum banana (micro-space-time). What shape does this quantum banana now have as a whole and in its parts? This infinitely reduced banana will probably have retained its familiar geometric shape regardless of its current size, so its body is non-Euclidean as a whole and still

Euclidean in its parts. According to experience, our universe could therefore be an infinite non-Euclidean and yet simultaneously finite Euclidean continuum.

The existing spatial structures [as with our quantum bananas (in distant observation) which can mentally grow into macrobananas (in close-up observation)] therefore repeat themselves independently of the variation in size. Size thus also appears to be relative like time, but inseparable from non-absolute time (Einstein's theory). A spatial structure remains a spatial structure regardless of its size and always has its own associated time from the fourth to the seventh dimension. By changing the magnification of a body, non-Euclidean space-time areas become visible in distant observation, but at the same time Euclidean space-time areas become visible in close-up observation. The banana can be seen curved in the distance and, when viewed up close, individual banana areas can appear non-curved, i.e. optically straight; of course, the (general-special) theory of relativity should be easier to understand with the help of this space-time reference.

So if our universe is an infinite non-Euclidean and at the same time finite Euclidean continuum, then the observation of spatial structures from a distance enlarging into proximity resembles Euclidean geometry and the observation of spatial structures from a distance, zooming out into the distance, resembles non-Euclidean geometry, an infinite sequence of natural spatial states in the physical sense with regard to the topologies of surfaces of bodies (as in our quantum banana) and for our universe (continuum) with regard to the topologies of patterns and structures of space-time states; the focus of the present consideration is the perceptible continuous change between non-Euclidean geometry to Euclidean geometry, from this Euclidean geometry back to non-Euclidean geometry, etc., because not only time appears relative, but also the space firmly connected with time, or rather its size. Space-time as a whole and in sub-areas thus always appears to be relative.

45 Gauss-Cartesian coordinates

In the general-special theory of relativity, Gaussian-Cartesian coordinates are used for supersymmetric and relativistic modelling and visualization of the movement of systems and bodies. For this purpose, a Gauss coordinate system is ascribed to an arbitrarily selectable space-time state, which is intended to represent a gravitational field and in which the individual bodies must therefore move freely but in a non-uniform curvilinear manner, while their Einsteinian mechanics simultaneously take place in a uniform rectilinear manner within a Cartesian coordinate system moving relative to this Gauss coordinate system.

The Gauss coordinate system is similar to Albert Einstein's gravitational field theory (2009) and this means that the movement of bodies is relatively non-uniform curvilinear within the space-time structure as in the general theory of relativity and at the same time relatively uniform rectilinear within the time-space structure as in the special theory of relativity. Accordingly, for this modern mechanics there is a relative compensation in four dimensions between the non-uniformity of a curved space-time and the uniformity of a rectilinear time-space, which should make it mathematically possible in the physical reality of our universe (continuum) to carry out a uniform rectilinear acceleration movement of bodies in gravitational field environments with a non-uniform curvilinear deflection of movement of these accelerated bodies.

To put it simply, Cartesian coordinates should make it possible to draw a straight line between an arbitrarily selectable reference system A and reference system B in order to be able to travel this distance uniformly with a test reference body (such as a spaceship) without the straight line becoming a geodesic during this journey due to differing gravitational fields in space and time. Therefore, the Cartesian coordinates are equal to the Gauss coordinates so that this *modern mechanics* can take place unaffected by non-uniform curvilinear motion influences.

In practice, in our solar system, the locally measured gravitational field differences should be taken into account for the Gaussian-Cartesian coordinate calculation. This can already be done with point masses (as with satellites), which can be placed at rest in five equilibrium points on an orbit of a body with respect to another body (gravitational field buoys), so that measurement data can always be up-to-date with regard to gravitational field shifts, for example due to changes in stellar activity and planetary movements.

46 The space-time continuum of general relativity theory as a non-Euclidean continuum

According to the general theory of relativity, our universe differs in its space-time structures in that our continuum does not (or no longer) resemble Euclid's geometry. There is therefore an infinite number of finite independent space-time regions. This number can have larger areas reduced to smaller curved areas (non-uniform non-linear micro-inertia systems) and at the same time smaller areas extended to larger curved areas (non-uniform non-linear macro-inertia systems) in several (here four) dimensions. This is all due to the field of gravity.

For more details, please refer again to section 15 and in general to the second section.

47 The space-time continuum of the special relativity theory as a Euclidean continuum

According to the special theory of relativity, our universe differs in its space-time structures in that our continuum (still) resembles Euclid's geometry. There is therefore an infinite number of finite independent space-time areas, this number can only have uniform rectilinear areas (homogeneous linear inertial systems) in four dimensions, none of this is due to the field of gravitation, because this objectified body is not contained in the special theory of relativity, but is only added in the general theory of relativity as a completion by Albert Einstein (2009).

For more details, please refer to section 14 and the third section in general.

48 The space-time continuum of general-special relativity theory as a non-Euclidean and Euclidean continuum

According to the general-special theory of relativity, our universe differs in its space-time structures in that our continuum does not resemble Euclid's geometry on the one hand and on the other. There is therefore an infinite number of finite, but not absolute, space-time areas. From four to seven dimensions, this number can only have non-uniform curvilinear areas (heterogeneous non-linear macro- and micro-inertial systems) within which there are uniform rectilinear areas of different sizes (homogeneous linear inertial systems). It is all due to the gravitational field, which must have a different influence on the geometry according to Euclid in the various space-time regions, for example due to heavy inertial stellar system masses that can be found locally nearby.

In order to easily understand the entire geometry of the general-special theory of relativity (matrix) with regard to our universe (continuum), it should have helped to have mentally read the second and then the third section, i.e. first the general and then the special theory of relativity, and to have visualized it. In this way, it should have been learned that non-Euclidean geometry appears completely impossible without Euclidean geometry, since it is a being and a wanting in our universe. Therefore, both theories of geometry have been deductively merged into the general-special theory of relativity and in which both ideas of geometry (theories) continue to be valid, since the fourth section represents an extension (completion) of the general and special to the general-special theory of relativity.

Accordingly, the general-special theory of relativity divides the universe (continuum) as a whole, like the general theory of relativity, into a non-uniform curvilinear area in which every line that is started should at some point find its beginning again due to the curvature of the universe (continuum), but which runs through many different non-uniform curvilinear areas due to the manifoldness of our universe (continuum), i.e. should have a different inward or outward curvature in different space-time areas. However, this variable curvature can decrease in individual space-time areas to such an extent that a uniformly straight course becomes conceivable over time until the angle of curvature increases again.

If, according to Einstein's theory, there is no absolute time and it only exists in connection with space, then one consequence is the following sentence: In areas without space there is also no time, and the opposite of this: In areas without time, there is also no space. This means that other general laws of nature could be contravariantly valid than those that apply covariantly in this universe (continuum). No space and therefore no time, or no time and therefore no space, appear to be conceivable with the general-special theory of relativity, especially in the multidimensional areas that could flow like transitions to differently curved areas as well as in event points of significantly ponderable masses, such physical states would now be conceivable. Various such mass event points could actually be connected via contravariant general laws of nature. This would also have to justify a uniformly distributed mass density with an average slightly different from zero, with which the general-special theory of relativity already stands up to a worldly experience.

According to the general-special theory of relativity, the speed of light movement must always be dependent on point coordinates, as at least one single large gravitational field (within) our universe (continuum) always exists. This existence of gravitational fields generally rules out a definition of space-time coordinates that is consistent with nature for a constant speed of light movement. This Euclidean mechanics is specifically possible again with the same definition of space-time coordinates, whereby a constant speed of light propagation can be described coincident with the law. In accordance with nature, the velocity movements of all bodies, such as the movement of light within gravitational fields, can thus be thought of in the general-special theory of relativity, i.e. transformed from a non-uniform (non-Euclidean) geometry of movement back into constant (Euclidean) laws of mechanics.

In the general-special theory of relativity, Einstein's gravitational field motion resembles a generalized, modern light field motion and thus a non-classical space-time motion. So if something like gravitation can objectively exist in a space as a field, then light can also exist in this space as a field, just like time, which, firmly connected to space, should also represent a field if Einstein's theory is generally followed. This time field with all t_{ik} (model tensor) could therefore oscillate (fluctuate) due to the gravitational field and light field like space itself, which can itself be understood geometrically as a three-dimensional field. The space-time concept here is geometrically similar to the field concept, whereby time, if its reference system is generally placed directly on the coordinate system of space, would also have to be three-dimensional. Space-time can therefore be viewed as an objectified body in the general-special theory of relativity with four dimensions (3 spatial coordinates + 1 time coordinate) or as an objectified field with seven dimensions (3 spatial coordinates + 3 time coordinates + 1 curvature angle coordinate). In both physical dimensioning cases, there is no absolute time as Albert Einstein (2009) also accepted from four dimensions. In the general-special theory of relativity, the curvature intensity of our universe (continuum) could be added as a seventh dimension due to the gravitational fields and light fields, but this should be easy to determine via the physically real π of our universe (continuum). Ultimately, the movement of time (and space) can be conceptualized in this way, which in Newtonian mechanics is absolute and therefore one-dimensional, uniformly rectilinear in one direction, but can now be considered non-uniformly curvilinear in this *modern mechanics*, i.e. more differentiated, so that time as a straight line could be closed at the end.

49 The exact representation of the general-special relativity principle

The basic understanding of the general-special principle of relativity is similar to the expression: *Every Gauss-Cartesian coordinate system is fundamentally equally suitable for representing the general laws of nature* (Einstein, 2009).

In general, this general-special relativity principle can also be expressed additionally with another theorem that psychologically arranges the same specific as its nature-coincident extension of the general relativity principle (Einstein, 2009). The equations describing the general laws of nature in general relativity must transition into nature-coincident equations by applying arbitrary substitutions of Gauss coordinates (x_1, x_2, x_3, x_4); because each (Lorentz) transition resembles a transfer from one Gauss reference frame to another Gauss coordinate system (Einstein, 2009). The general laws of nature are represented in the special theory of relativity with equations that are transformed into equations that are consistent with nature when new spatial and temporal coordinates (x', y', z', t') of another (Galilean) reference system K' are generally determined instead of spatial and temporal coordinates (x, y, z, t) of a Galilean reference system K by the Lorentz transformation (Einstein, 2009).

If, in general, an experiential perception with three dimensions must be maintained for both space and time, then a psychological development, which we understand on the basis of the elaborated principle of general-special relativity theory, can be described as a theory as follows (Einstein, 2009): Practically rigid reference bodies do not exist within gravitational fields and therefore in this gravitational field also do not exist with a uniform geometrical constitution according to Euclid; therefore a notion of (practically) rigid reference frames does not work in general relativity theory (Einstein, 2009).

Building on Albert Einstein (2009), the marble table surface and the flame, which leads to a heat-induced deformation (contraction) of the square construction design of the universe (continuum) according to Euclid, should be recalled at this point.

Non-rigid reference systems are therefore generally introduced, which move arbitrarily and can experience arbitrary changes in their construction due to the gravitational field (Einstein, 2009). This non-rigid reference system resembles an arbitrarily selected Gauss coordinate system with three dimensions (Einstein, 2009).

Gravitational fields can have an influence on the spatial and temporal movements of quasi-spherical systems and bodies accelerated therein, i.e. also on the scales and clocks (or rather on their physical sense). so that the scales experience different degrees of contraction depending on the mass field distance and the homogeneous-linear definition of the time that really exists in the universe found directly with clocks justifiably does not reach the confirmation distance on the clock face in general-special and general relativity theory compared to special relativity theory, in which, however, no light field and no gravitational field acts on the time machines (clocks) (Einstein, 2009). For the determination of time, clocks are used which are subject to an arbitrary, strongly deviating non-uniform law of motion, which are generally fixed (at rest) at every point of a non-rigid reference system, and which have only one physical sense, to display the metrics (quantities) of (also) locally stored clock neighbors (infinitely often) at the same time (Einstein, 2009).

A special case of gravitational fields exists relative to a non-Galilean (Cartesian) reference frame (Sections 8 and 11) (Einstein, 2009). Certain assumptions allow the non-Galilean (Cartesian) reference frames to be connected to Galilean domains (Einstein, 2009). In the universe, Galilean domains exist in the special theory of relativity, i.e. only those within which a gravitational field does not exist (Einstein, 2009). A Galilean reference system (geometry system) is selected, i.e. a practically rigid system with a mechanical state selected in accordance with nature, so that relative to this system the Galilean expression regarding a uniform rectilinear direction of motion of "independent" test reference bodies (point masses) conceivably continues to exist (Einstein, 2009).

Figuratively speaking, the general-special principle of relativity can also be understood in this way: Between two bodies, a uniform rectilinear motion in a certain direction should be possible with a sample point mass independent of these bodies (for example with a spaceship). However, since all bodies (should) move according to classical mechanics, the distance between the two bodies and the angle of the imaginary straight line between them are constantly changing. According to Einstein's mechanics, there are further influences observed from outside, at least the two gravitational fields of the quasi-spherical reference bodies (and all gravitational fields of neighboring masses in the universe), which should constantly lead (without corresponding continuous correction) to the straight line becoming a geodesic and the test reference mass moving in a non-uniform curvilinear manner. This means that this non-rigid point mass (for example an asteroid) should accelerate faster or slower and change its direction of motion (mechanical deflection) depending on the nearby gravitational fields.

A journey as an example in our solar system between Earth and Mars could thus be designed more accurately and the gravitational light field influence (anti-gravitational field influence) could be calculated and experienced, since from a certain point on the imaginary straight line the Martian gravity should become stronger than the retreating Earth gravity (anti-gravitational field). This should be the physical case from a certain distance Earth-Mars, the movement would thus remain uniform, since depending on the desired cruising speed v an acceleration adjustment can take place, here a speed reduction due to the stronger Martian gravity, and rectilinear because the curvilinear gravitational light field effect of the bodies concerned can be constantly compensated by means of the approach angle.

If we weigh a spherical body of iron on a planet such as Earth and obtain a weight of 1 kilogram, for example, and place this sphere on another planet such as Mars purely in our minds, then we can ask ourselves the following: What weight does the sphere have in a different gravitational light field without any theory being known to me? In general, at the beginning of the consideration, we should think that the sphere should have a lower weight, especially in the case of Mars, because its gravitational light field should be proportionally eccentrically smaller than that of the Earth, and if we were to take our weighing machine from the Earth to Mars, then we should actually notice a difference in mass gravity. But from a physical point of view, is this a difference in weight? What could really have happened to the iron ball instead, especially because, as I said, according to the theory, no constant for gravity comes to mind? The sphere on Mars could have undergone a minimal expansion in all directions, so it has become a very small bit larger than on Earth. However, this shift in size is not visually significant, because according to the general-special principle of relativity, weightlessness of masses could exist anywhere in the universe (anti-gravitational field), even on planets, which is why Albert Einstein (2009) could have equated mass gravity with mass inertia. So only inertia could really exist physically in the universe, which means that there could be no weight, instead only a velocity in the direction of the center of mass of inertial reference systems and inertial reference bodies. Even now, the classical gravitational constant according to Newton would no longer be tenable (i.e. wrong in principle), which should make the desire for a modern gravitational variable noticeable in relation to the universe. Since we generally only have our perception as a form of geometry due to the relativity of space and time and we are familiar with the optics of both planets, we can say according to these observed dynamics of optics: π i.e. circumference U divided by diameter d of the great circle (equator) of the body should be useful in this derivation of a modern gravitational variable. For diameter d, we can of course additionally take into account the angle of curvature α due to the gravitational light field starting from the circumference in a three-dimensional spatial view of a spherical reference body. The covariant system of equations with respect to these two snapshots from the same distance is therefore $g_{ik} = \pi_{Earth} = \pi_{Mars}$ or $g_{ik} = U/d_{m1} = U/d_{m2}$, but according to this theory of everything, a more advanced π could be even more precisely suitable for the dark quasi-spherical universe, which I will not go into in more detail due to the necessary mathematics.

According to these optical dynamics, we can now apply an arbitrarily chosen scale to the two snapshots and perform a differentiation approximating the modern gravitational variable according to one of the two terms of $d_{m1}d_{m2} = U_{m2}/U_{m1}$, which should express a dependence with respect to the modern variable weight force for the gravitational light field g_{ik}, which should truly represent an inertial force radius $r_{ik} = g_{ik} / 2$ to the center of mass of a body. If we consider the root in the latter term, this results in a gravitational variable $k_{ik} = \sqrt{(g_{ik} / 2)}$ based on Gauss, which could enable us to make more generalized physical mathematical statements regarding gravitational light fields without an assumed mass gravity as according to Isaac Newton only on the basis of mass inertia as according to Albert Einstein (2009).

If the described distance between the two bodies is now thought of in an intergalactic scale, after the light deflection of the targeted star is clear and thus its physically real position is determined at the same time (Section 3.2), it is generally noted that many disturbing nearby gravitational fields can exist, which can lead to a deviation of the travel route. Whether this journey through our quasi-spherical space-time solar system is possible at all would then turn out to be covariant or contravariant according to nature, because at the outermost gravitational light field surface, which could look like a practically invisible but transparent mirror wall if space spheres in general were actually closed following the field geometry of light gravity, an infinitely manifold curvature (i.e. an unimaginable dimensioning of dark substance) would have to be perceptible like an impenetrable boundary horizon; In this physically general-special case, multidimensional manifold space-time twists between the space spheres would be conceivable mass-centered in our universe. Without the general-special theory of relativity, such a movement should therefore end in nowhere in multidimensional space-time without calculating the gravitational field deflection. Since the space-time structures also (should) move between intergalactic bodies, this could mean that as soon as a flight is made into a correspondingly higher curved space-time area on which various gravitational fields have an influence, unforeseen point events such as collisions with another celestial body, such as a deflection into a black hole or onto a free-flying moon, could otherwise not be ruled out, especially if the solution to the speed of light problem is conceivable on the basis of the principle of general-special relativity.

50 The solution to the problem of the speed of light based on the general-special relativity principle

Until now, scientists and others have wrongly assumed that the speed of light c represents a limit in the physical sense of a maximum speed v of bodies in a vacuum. This false assumption can be traced back to the law of the constancy of the speed of light c, which for Albert Einstein (2009) with regard to the derivation of the special theory of relativity only served as a thought example for the description of the uniformity of a movement with a constant speed v. The Lorentz factor, which represents a mathematical limiting function for the movement of light, was therefore also adopted into the special theory of relativity with the Lorentz transformation. Other conceivable influences such as friction and the gravitational light field are not taken into account here, as they influence the speed v but cannot lead to such a mathematical maximum function (system of equations) in a way that is consistent with nature.

Fizeau's experiment (Section 33) was also no reason for Albert Einstein (2009) to assume that the law of constancy of the speed of light c represents a maximum property of the speed of light w. This is because Albert Einstein (2009) used his (special) theory of relativity to determine the velocity v of the fluid flow with influence on the movement of light w with his equation for parallel velocities $W = (v + w) / (1 + vw/c^2)$. This is a specific speed of light w which is determined experimentally and mathematically relative to a tube filled with liquid (on earth). Fizeau's experiment thus proves Albert Einstein's theory of relativity in an exact manner, but has no statement about a maximum speed of light.

Experience has shown that no test reference body has ever flown at (approximately) the speed of light c in practice. However, this is not proof of a speed limit according to the law of constancy of the speed of light c, but merely an important indication that our understanding and observation of the speed of bodies may not yet have been sufficient (speed of light paradox). This consideration regarding spatial reference bodies that are driven too slowly is (also) to be understood independently of a practically infinite speed v of light in physical reality, because this must (nevertheless) be maintained for all moving bodies in our universe according to the general-special theory of relativity as a practically infinite potential function (system of equations) for determining a multidimensional surface displacement of the test reference body that depends on the directional speed, analogous to light.

The speed of light paradox is solved in the general-special theory of relativity by the fact that instead of a real observation of a freely moving spatial reference body, an understanding of this observation is simply formed, in accordance with the general laws of nature or experience.

The speed of light paradox already falls apart when the following theoretical possibility is considered: If a star accelerated (approximately) to the speed of light w = c is considered, then, according to the law of constancy of the speed of light c, this star radiates light bodies in each of its directions at the speed v = c; for the light radiation emanating from this star, the following then applies at a rectified speed W = v + w = c + c = 2c; consequently, according to the law of addition, the light rays are already moving at the speed W = 2c. This renders the law on the constancy of the speed of light c invalid in its physical sense, because the star moves at c and the light at 2c, but the distance between the two is always given by the relative speed c, whereby the previous law can be renewed to the law on the simultaneous constancy of the relative speed of light c with regard to bodies moving in the same direction.

The law of the constancy of the speed of light c also becomes invalid from the point of view of an observer on a rotating moving body who observes this star and the light at the same time. An observer always sees the star moving at c and the light moving at 2c in the same direction. An observer must always be able to see a constant field range in every direction of a reference body emitting light, which means that if the obsolete law were really physically valid, everything would have to be dark when viewed in the same direction of the star's movement, but it won't be, instead it will also be bright there, as the light in a vacuum should not be exposed to any friction but to the gravitational field of its own star; However, if this gravitational field were too strong for the light to develop a sufficiently high escape velocity v, then no light would be visible around the star and the star would appear black like a multidimensional hole, around which dark heat radiation would be visible.

The question of the existence of a practically finite speed of light c or a practically infinite speed of light v has not yet been fully answered. The speed of light was measured in the case of the moons of Jupiter in 1676 by the astronomer and mathematician Ole Christensen RØMER, who believed that the speed of light c was practically finite; I made the same observation with a Dobson-Newton telescope with a 200 mm aperture, but made a different assumption, namely that the speed of light v is practically infinite.

If I assume a practically infinite speed of light v to answer the previous question of existence on the basis of the general-special theory of relativity, the light of the respective moon of Jupiter would immediately be visible here on Earth again as soon as this moon emerged from the shadow of Jupiter. Instead of an absolute time, as was apparently already common before Newtonian mechanics, because this is already prone to error due to the mechanical clock movements alone and thus deviations in the experimental results become conceivable, I also assume a time that is firmly connected to space (Einstein's theory). If I stand on Earth (x, y, z, t) as an observer of Jupiter's moons (x', y', z', t') equipped with a clock and a telescope and write down every time event (t = t') from which I can see the respective Jupiter moon emerge from the shadow. Then I only write down moments of simultaneity of the movement of bodies that I have observed on one of the two bodies, which means that at these event points the Earth's time appears to be the same as Jupiter's moon time (t = t') due to the movement of light.

If the distance between Earth and Jupiter's moon remained unchanged at these time event points (t = t'), no statement could be made about the question of the speed of light. Instead, a variable distance must always be conceivable between each of the space-time points, because the light reflected back from Jupiter's moon to Earth always moves along a variable multidimensional geodesic due to the nearby gravitational fields of the Earth, the Sun and the other celestial bodies.

So does the light in our universe move as an objectified body either as a single beam of light or as a common field of light in space? Observed from the surface of Jupiter's moon, it will be a large common field of light propagating towards Earth within a multidimensional geodesic, and as soon as the light appears on Earth it will appear to the observer there as a small single beam of light. The variable multidimensional geodesic is therefore a curved cone whose tip appears to be connected to the absorbing body, i.e. a purely optical curvature singularity of light. This would make it possible to imagine a certain, practically finite range of light S, but still no statement about a finite speed of light c. This is because instead of moving in a tube filled with resting liquid (Fizeau experiment), the light now moves in a space filled with resting gravity (analogous to a field).

The addition law of parallel velocities $W = (v + w) / (1 + vw/c^2)$ is similar to experience because it was confirmed by Fizeau's experiment (Section 33), which illustrates the movement of light in a fluid at rest (Einstein, 2009). The question in this context is: How fast does the light from a moon of Jupiter move with a velocity w in a space K in the direction of the Earth if the space K is traversed by a practically stationary gravitational field K' with the velocity v (Einstein, 2009)? Regardless of whether the gravitational field K' is moving relative to space K and the light or not, the movement of light relative to the gravitational field must always take place at the speed w (Einstein, 2009). The speed of light relative to space K must be determined because the speed of light relative to the gravitational field and the speed of the gravitational field relative to space K are given (Einstein, 2009). The velocity W describes the movement of light relative to space K (Einstein, 2009).

A direct, present remote effect of gravitation without the field concept according to Newton's theory does not exist in the (general-special) theory of relativity, therefore this momentary remote effect with practically infinite speed of motion v is always exchanged by the remote effect of the gravitational field propagated with the practically finite speed of light c (Einstein, 2009). The theory of addition with regard to parallel velocities of light moving in a space (field) filled with stationary gravity is therefore: $W = (c + w) / (1 + cw/c^2) = (c + w) / (1 + w/c) = (c + w) / (1 + w/1 \times 1/c) = c / (1 + 1/c) = cc / (1 + 1) = c^2 / 2$. The speed of light w thus always appears to be proportionally dependent on the speed of the space (field) filled with resting gravity, which means that no practically finite speed of light c can be determined due to the lack of mass gravity [equal to mass inertia according to Albert Einstein (2009)]. A practically infinite speed of light w therefore appears to be a more attainable truth in this trans-light theory according to experience.

It is therefore not possible for me to draw conclusions about a practically finite speed of light c due to the different spatial distances (x, y, z = x', y', z'), which change over the course of a calendar year, especially as the deflection of light towards our sun is not taken into account in a way that is consistent with nature due to its gravitational field (Section 3.2), just like all other gravitational fields in our solar system or universe.

New views on the entire universe

51 Limitations of Einstein's relativity theory with regard to the universe

The general theory of relativity adopts the law of the constancy of the speed of light c from the special theory of relativity, which could have resulted in mental limitations with regard to the general laws of the nature of our universe in direct connection with the assumptions regarding gravitational fields existing in space. Of course, these limitations of Albert Einstein's statements (2009) are minimal and formulated by him as challenges, such as that this field relativity theory is not yet completed, therefore appears solvable by a subsequent derivation of the covariant systems of equations – here to the general-special theory of relativity.

The following restrictions apply with regard to the covariant general laws of nature. Gravitation exists as a field in space and has an influence on light in space in the general theory of relativity. However, this gravitational field theory only says something about the color shift and directional deflection of light and nothing about a practically finite or infinite speed of light c or v, except that in a special case of this theory the law of constancy of the speed of light c is integrated. A concept of speed with regard to light is only possible indirectly via the gravitational field. For if the existing speed of attraction of the ponderable mass is greater than the necessary speed of repulsion of light, then the speed of light appears to be finite to a certain extent within a practically infinite acceleration metric, but a constant field speed of gravity and light cannot be experienced as a result; instead, a multiplicity of the universe and its natural metrics for bodies and their interactions is pointed out, whereby their speed appears to be variable at all times. The speed of light is therefore directly dependent on the speed of the gravitational field or both are equated in a covariant system of equations. A practically infinite acceleration metric is therefore generally conceivable, in which the velocity of bodies (including light) appears to be finite due to the gravitational field, but this finite concept must be interpreted in such a way that this consideration can assume practically infinite forms within the acceleration metric.

The covariant general laws of nature must always apply in full with regard to the gravitational field theory derived by Albert Einstein (2009), which means that an observable deviation would represent a contradiction and require a modification of the gravitational field theory. However, observations can show that the gravitational field laws always resemble the physical case described, and this seems to apply to gases in particular, as well as to other matter. This theory is now important for planetary research, for example, especially with regard to gaseous planets such as Jupiter and Saturn. In general, for gases that are not at rest, such as those that exist in our planet's moving atmosphere, a cold gas flowing downwards is more inert (heavier) than a warm gas flowing upwards. The flow speed of gases on bodies such as planets can be due to the speed of the gravitational field coming from inside and the speed of the light field coming from outside. When it reaches the bottom of the body illuminated by light (e.g. a planet) and is (partially) reflected back, the light heats the gas during this movement, if present it rises, and at the same time the gravitational field pulls the colder gas back down to this open local point. This results in friction between the gas particles during the flow movement and an atmospheric equalization on the surface with regard to temperature, mass and speed, i.e. according to the natural coincident expression for the energy $E = Tmv^2$ according to the general-special theory of relativity; a gravitational field can thus also be assigned a certain ordering function in our universe (continuum).

The ordering function of the gravitational field can already be experienced if the two poles of such quasi-spherical reference bodies (e.g. a planet) are generally considered. This is because, in comparison with their rotating equator and the resulting differential rotation of the atmospheric gas, these are generally confronted with more or less light at half the orbital curve, which means that the gravitational field should have a greater local significance there, and not just for atmospheric equalization; electrodynamic effects on the magnetic field of the bodies are considered in the next paragraph. Therefore, the warm gas usually sinks down cooled over the poles and spreads out on the ground until this gas rises upwards again, predominantly warmed up at the equator at the latest; a gas particle wind thus always arises unevenly swirled but simultaneously caused by the *modern mechanics* of the two superimposed fields of light and gravitation. Nature consistent with the observation that heated gases can rise, it therefore does not seem possible that covariantly determined general laws of nature can contravariantly circumvent the general theory of relativity – or could not yet or only incompletely be recorded. So why the warmer, lighter, less inert gas particles can flow upwards purely optically unaffected by the gravitational field, while the colder, heavier and therefore more inert particles of matter must always be directed downwards towards the surface of the body in an orderly manner by the gravitational field, is described by Einstein's gravitational field theory and therefore this nevertheless always appears to be tenable, a limitation of the general laws of nature only takes place (as previously derived) with regard to the procedure for describing the speeds of movement of ponderable particles of matter.

The gravitational field must have electrodynamic effects on the magnetic field of the bodies (for example of a planet) or be able to determine this electromagnetic field according to Einstein's theory. If I imagine a quasi-spherical moving rotating body that is heavy, i.e. inertial, and localized in an airless space (vacuum), then according to Newtonian mechanics there should already be a proportionality between the geometry of the body and its electromagnetic field in the physical sense that is generally known from iron filings, which stick to a roundish magnet in a corresponding shape and, figuratively speaking, stand up like the spines of a hedgehog in the space-time of our universe (continuum). Depending on the rectified speed of movement W and the gravity G of two bodies and their distance r, the force of the electromagnetic field can be determined using the expression: $W \times G_1G_2/r^2$. The electromagnetic field force e_{ik} appears to be located on a (curved) path between two gravitational field aptitude points and to be proportional to the product of the two gravitational fields and inversely proportional to the square of their distance. Due to the derivation of the gravitational field laws by Albert Einstein (2009), a symmetrical property was also assumed here with regard to the electromagnetic fields. Einstein's gravitational field theory therefore appears to conform to the covariant general laws of nature, although this still leaves some scope for model design for other related (contravariant) general laws of nature.

The following restrictions apply to contravariant general laws of nature. With regard to the general theory of relativity, there are no restrictions for contravariant general laws of nature, as these are not taken into account by the theory due to purely covariant systems of equations. However, contravariant general laws of nature can still be related to covariant general laws of nature, especially since the universe (continuum) turns out to be non-Euclidean as well as Euclidean, which are already described in the general theory of relativity, for example via the Lorentz coordinate transition, since their components must generally be transformed uniformly and rectilinearly. A mathematical application of the Lorentz transformation therefore requires the assumption of covariants. Contravariant general laws of nature, on the other hand, appear to be more difficult to transform, as they are non-uniform and curvilinear. This does not correspond to the principle of equivalence practiced by Albert Einstein (2009), so contravariant systems of equations should not simply be equated with covariant expressions. If I imagine two widely separated space-time areas in which only covariant laws of nature apply in general, i.e. Einstein's gravitational field laws, the gravitational field concentrated on one event point each could lead, depending on the mass inertia (equal to mass gravity) and its rotational speed, to a vortex forming from two sides and combining to form a new, completely swirled space-time. This altered space-time would therefore have to be viewed differently from the other covariant components of the universe, as other contravariant general laws of nature could apply in this multidimensional region. Viewed from the outside by an observer, such a vortex phenomenon could alternately shimmer brightly, as the light from the stars could be deflected and constantly change direction. If a body were to enter such a diversely swirling, probably bouncing vortex, its direction of movement could depend on the larger gravitational field event point. If this body is able to withstand the transition into this gravitational field movement at all, if so, the tidal accelerations prevailing there could not stretch it first, but tear it apart into its particles immediately. For contravariant general laws of nature, individual relativity models that have been newly designed or modified for these physical scenarios must therefore always be set up in a contravariant manner, the content of which can, however, be linked to the existing covariant (general) theory of relativity at individual transitions.

52 The theory of an infinite and yet finite universe

If I follow the idea of Albert Einstein (2009) (Section 19) that the universe represents a sphere of space or probably a space ellipsoid, which appears to be infinite, but which could nevertheless have a barrier (boundary) to the outside inside due to the gravitational field, which is probably brighter than the space in front of it, but within which a practically finite number of gravitational field-related quasi-spheres (space spheres) could exist, all of which would have to be quasi-symmetrically curved on the inside and therefore have boundaries (transitions) and could overlap (transform) several times in their dimensions towards the outside, right up to the edge of the space ellipsoid, a manifold elliptical space-time continuum (universe) would have to exist according to the geometry. There could therefore be closed spaces (or space fields) such as the (quasi-)space spheres, which nevertheless have no visible boundaries (barriers). In the same way, there could be open spaces such as the space ellipsoids, which nevertheless make visible boundaries (barriers) observable at the same time. The higher (or lower) the curvature in the (quasi-)space sphere, the smaller (or larger) the (quasi-)space sphere could be (and vice versa). A high finite number of small (or large) quasi-spheres of space would mean a practically infinite (or finite) curvature for the space ellipsoid itself, whereby our universe (continuum) could become warmer (or colder) towards its edge; this is all conceivable due to the geometrically possible multiplicity of space spheres and space ellipsoids, which could also determine a truly natural thermodynamics of our universe (continuum).

As a consequence of this theory of an infinite and yet finite universe (continuum) or its entropy, an awakening idea could be noted: is our universe dark inside (except) at practically finite event points, although infinitely open with a surface horizon consisting of limiting (insurmountable) temperature point events, but nevertheless finite due to the gravitational field-related (quasi-)space spheres (space-time realities) surrounding us? According to the general and general-special theory of relativity, this could indeed be the case, but space-time practice has not yet been able to gain sufficient experience of this. People are therefore becoming more and more aware of their solar system or their (quasi-)space sphere and thus of the entire space-time elliptical universe (continuum); this theory of everything, which I call the Einstein-Hawking-Theory, could therefore help to clarify for our present whether and to what extent all bodies could really be physically exposed to the general laws of nature of an infinitely (large) and yet finitely (tiny) extended black space ellipsoid (universe).

53 The spatial structure according to the general-special relativity theory

Since, according to the general theory of relativity, a universe according to Euclid could only exist at individual coordinates with condensed matter, the average density of these only locally distributed masses would have to be zero at every event point in this continuum (Einstein, 2009). However, our universe has an average mass density slightly different from zero, which makes a Euclidean universe inconceivable for this physical case (Einstein, 2009). The average mass density confirms that our universe should exist as a space sphere (or space ellipsoid) with a uniformly distributed mass (Einstein, 2009). We speak of a space ellipsoid because matter in physical reality is distributed asymmetrically in many space spheres, whose behavior is thus also dimensionally asymmetric, which means that our universe is deformed into a quasi-space sphere (a space ellipsoid) due to the gravitational field, which is why our continuum should not be infinitely large (Einstein, 2009). The general theory of relativity provides the "radius" R as a result of the equation $R^2 = 2/kp$ (Einstein, 2009) via the ratio of the expansion of space to its average matter density. Readers can, like Albert Einstein, assume 1.08×10^{27} for 2/k and multiply this value by p (average mass density) and finally take the square root, then they should have determined an unexpectedly very small "radius" R, depending on which value is used for p (Einstein, 2009). If everything that is in our universe according to the theory of everything should appear very large based on the observation of our reality, then everything in the physical reality of our continuum could be very small at the same time (Einstein-Hawking-Theory).

Since the answer, which was clarified in the previous section, to the question is "everything according to theory" in a black hole perhaps (i.e. no longer clearly no and today also not definitely yes) (Section 51), it can be expressed with regard to the spatial structure that the spatial structure according to the general-special theory of relativity – compared to the spatial structure in a black hole – in our universe (continuum) could be uniformly (spatial sphere) or non-uniformly (spatial ellipsoid) structured. For example, the assumption from the general-special theory of relativity alone that light could also behave like a field during its movement, which could appear optically like a curvature singularity radiated from the surface of the body, whose starting points adopt this body geometry for the light field and whose end points with increasing distance could generally (without gravitational field shift) make this radiated light field surface appear more and more adjacent (condensed) at each event point, points to a non-uniform (non-Euclidean) motion on a seven-dimensional geodesic field line between the luminous and illuminated inert (heavy) bodies. Such a curvature singularity is therefore to be considered seven-dimensional in the general-special theory of relativity, since its description is possible via three spatial coordinates and three time coordinates as well as a bending angle coordinate. This should be able to (mathematically) express their gravitational field-related superimposed manifold (multiplicity), especially in the physical case of the light field and the space ellipsoid, each viewed from their surfaces, because a practically infinite manifold of our universe (continuum) could be represented by finite field line-like curvature singularities rooted in each other. This light field could become practically infinitely non-uniform (manifold) due to the distance to the different gravitational fields of the bodies, because a light field that is emitted by an inert (heavy) body as a quasi-spherical light surface could, when it approaches a lighter (or heavier) body, become less uniform. (or heavier) body, it could decrease (or increase) in accordance with π; this illustrates an analogy with regard to a practically infinite spatiotemporal universe with event boundaries (space ellipsoid), which could nevertheless have finite temporal worlds without event boundaries (space spheres). This means that our world could be one of the worlds in the world that could also exist separately from the other worlds in the world. Parallel movements of space spheres within a space ellipsoid could then represent either gravitational field dynamics or reflections (deflections) of their own light field, i.e. in both physical cases transformations (transitions) in space-time, because "everything" in our practically infinite universe appears "according to theory" to be non-uniform (non-Euclidean) and yet uniform (Euclidean) in finite areas (Einstein-Hawking-Theory).

Fifth section: Appendix

54 Advanced derivation (deduction) of the transformation according to Lorentz

Now we know that there really could be no physical limit to the speed of light. However, the question arises as to whether this could not only apply to the light field, i.e. also to inert (heavy) bodies in our universe? All that is required is to modify the Lorentz transformation (72 and 75) so that it represents the realization c = v = w, i.e. the constancy of the relative speed of light c for two simultaneously considered body reference systems. The velocity w applies to the moving spatial state and the velocity v to the arbitrarily selected test reference body. Any point mass that is not at rest at the same time as the moving spatial state can be selected as a test reference body.

(72) $x' = \frac{x-wt}{\sqrt{1-\frac{w^2}{v^2}}}$ (Modified Lorentz transformation)

(73) $y' = y$ (Einstein, 2009)

(74) $z' = z$ (Einstein, 2009)

(75) $t' = \frac{t-\frac{w}{v^2}x}{\sqrt{1-\frac{w^2}{v^2}}}$ (Modified Lorentz transformation)

But which objectified state represents the moving space? If, for example, I consider the space that a matter occupies at a point, its spatial state is always in constant motion at the same time as the body propagates. Does this small moving spatial component not relate to a comprehensive multidimensional spatial area that contains this direction of movement? According to Albert Einstein's general theory of relativity (2009), this small spatial component moves in a non-uniform curvilinear manner in the gravitational field of our infinitely large universe. The space component will therefore be accelerated faster, equally or slower, simultaneously observed to the gravitational space field of our universe. In Newtonian mechanics, all objectified bodies are in motion and time is absolute, so this also applies to both spatial states.

However, if there were no motion in all bodies in the universe for some as yet unthinkable reason and time, as Albert Einstein (2009) said, is not absolute but firmly linked to the state of space, then every motion would represent a punctual change of space-time in several dimensions, whereby, according to the new law of the simultaneous constancy of the relative speed of light c, all body movements would occur simultaneously. A sequence of movements would therefore be a kind of illusion based on our absolute observation of time while the point of matter is moving. In this physical sense, the reflections are not an illusion, but simply different space-time states to be seen, which would either exist one after the other in absolute time or could all exist simultaneously, since time is not absolute. So what reason can there be that we cannot see this simultaneity of movement of bodies (space-time states)? There could be an infinite number of space-time states of motion whose individual reflections should be dependent on the velocity v, the higher the velocity v, the smaller the moment, i.e. the shorter the moment of time t that can be seen.

One possible conclusion from this experience is that the speed of bodies is actually much higher in physical reality than we believe or can measure today. Then it would only appear optically as if there were a real movement and not an infinite number of individual space-time states whose reflections could only be seen for a very short time and then immediately disappear again. Our brain could falsely run this from a starting point to an end point like a reality movie in several dimensions; as a result, the apple that Newton allegedly fell on his head while sitting under a tree would never have fallen, but would still be hanging on the tree, but in a different space-time mirror dimension. The falling of the apple would then be easily explained with the Lorentz transformation and the general-special theory of relativity; this is a "*modern mechanics*" integrated with a minimum of 11 or 12 dimensions or more differentiated with a maximum of 18 or 21 dimensions for two non-absolute space-time states with four or seven dimensions each (three space coordinates (x, y and z) each). seven dimensions (three space coordinates each (x, y, z), a time coordinate t optionally with regard to x, y and z, the latter "reflecting" past and future simultaneously in the present, and optionally a curvature angle coordinate α) and a covariant space transition with the three space dimensions and the curvature dimension between each non-absolute space-time state; in a contravariant system of equations, a dimension t separate from the covariant time is to be added to the space-time transition optionally with respect to x, y, and z. It is therefore conceivable that the moment in which the apple is still hanging on the tree still exists, just as the moment exists in which the apple fell on Newton's head or collided with his body.

At this point, I have contradicted Albert Einstein (2009) somewhat with regard to simultaneity and his therefore only once existing time coordinate t, which is why Albert Einstein assumes 10 dimensions (9 of which are spatial coordinates) for our universe in his (general) theory of relativity, which is also true if all movements in space according to Newton should take place in a non-absolute time according to Albert Einstein (2009), according to which time mechanics should only appear possible in one direction. In my general-special theory of relativity, on the other hand, as a consequence of the general and special theory of relativity of Albert Einstein (2009), I postulate that there could not really be any movements according to Newton's physics and that a non-absolute time according to Albert Einstein (2009) is also established, but I consider each individual space-time state as a single movement according to the principle of equality (equivalence principle); these could be separate (deceptive) space-time reflections (instead of space-time movements) of bodies, whereby I explicitly contradict Isaac Newton's theory in a differentiating way and at the same time retain Albert Einstein's theory in an integrating way. Thus, no moving spatial state could exist with the velocity w, from which the Lorentz transformation for the description of space-time dimensions should follow; here v stands for the number of possible reflections between the start and end of an observed, only apparently existing motion of a body, whereby only in the physical case of a contravariant transition should a time mechanics in two different directions appear conceivable:

(76) $x' = \frac{x-t}{\sqrt{1-\frac{1}{v^2}}}$ (Modified Lorentz transformation for the representation of space-time state reflections)

(77) $y' = y$ (Einstein, 2009)

(78) $z' = z$ (Einstein, 2009)

(79) $t' = \frac{t-\frac{1}{v^2}x}{\sqrt{1-\frac{1}{v^2}}}$ (Modified Lorentz transformation for the representation of space-time state reflections)

55 The Minkowski universe with six dimensions

The next simplified assumptions are based on Minkowski in accordance with the Lorentz transformation (Einstein, 2009). In this universe (continuum) with six dimensions (Section 48), there are two point-event neighbors to be investigated, to which a mutual position is attributed with respect to the Galilean reference system K according to the differences in the space coordinates *dx*, *dy* and *dz* and the differences in the time coordinates *dt(x)*, *dt(y)* and *dt(z)* (Einstein, 2009). The same differences exist with regard to the second Galilean reference system K' for the two point events *dx'*, *dy'*, *dz'* and *dt(x')*, *dt(y')*, *dt(z')* (Einstein, 2009). Accordingly, the following relation system (80) is always valid between the two point event neighbors (Einstein, 2009).

The derived equations for c = v relevant for these space-time coordinates are $x + y + z + \sqrt{-1}vt(x) + \sqrt{-1}vt(y) + \sqrt{-1}vt(z) = x' + y' + z' + \sqrt{-1}vt(x') + \sqrt{-1}vt(y') + \sqrt{-1}vt(z')$ and $x_1^2 + x_2^2 + x_3^2 + x_4^2 + x_5^2 + x_6^2 = x'^2_1 + x'^2_2 + x'^2_3 + x'^2_4 + x'^2_5 + x'^2_6$; these equations are also valid with respect to coordinate differences, i.e. also with respect to infinitely small coordinate differences (coordinate differentials) (Einstein, 2009).

(80) $dx^2 - c^2dt(x)^2 + dy^2 - c^2dt(y)^2 + dz^2 - c^2dt(z)^2 = dx'^2 - c^2dt(x')^2 + dy'^2 - c^2dt(y')^2 + dz'^2 - c^2dt(z')^2$ (Einstein, 2009)

The relation (80) is similar to the reason regarding the (still existing) validity of the Lorentz transformation (Einstein, 2009). I formulate the following sentence: The number ds^2 that is assigned to both point neighbors in the space-time of our universe with six dimensions is the same in every Galilean reference system (81) (Einstein, 2009).

(81) $ds^2 = dx^2 - c^2dt(x)^2 + dy^2 - c^2dt(y)^2 + dz^2 - c^2dt(z)^2$ AND $ds'^2 = dx'^2 - c^2dt(x')^2 + dy'^2 - c^2dt(y')^2 + dz'^2 - c^2dt(z')^2$ (Einstein, 2009)

If x, y, z, $\sqrt{-1}vt(x)$, $\sqrt{-1}vt(y)$, and $\sqrt{-1}vt(z)$ are generally exchanged for x_1, x_2, x_3, x_4, x_5, and x_6, then the general result is also that the number ds^2 is not dependent on the choice of reference system (82) (Einstein, 2009).

(82) $ds^2 = dx_1^2 + dx_2^2 + dx_3^2 + dx_4^2 + dx_5^2 + dx_6^2$ AND $ds'^2 = dx'^2_1 + dx'^2_2 + dx'^2_3 + dx'^2_4 + dx'^2_5 + dx'^2_6$ (Einstein, 2009)

We denote the number ds as the distance between the two point events on six dimensions (Einstein, 2009).

The derived constructs are as follows $dx_4 = -1v^2dt(x)^2$ and $dx'_4 = -1v^2dt(x')^2$ and $dx_5 = -1v^2dt(y)^2$ and $dx'_5 = -1v^2dt(y')^2$ and $dx_6 = -1v^2dt(z)^2$ and $dx'_6 = -1v^2dt(z')^2$ (Einstein, 2009).

If the imaginary model coordinates $x_4 = \sqrt{-1}vt(x)$ and $x'_4 = \sqrt{-1}vt(x')$ and $x_5 = \sqrt{-1}vt(y)$ and $x'_5 = \sqrt{-1}vt(y')$ and $x_6 = \sqrt{-1}vt(z)$ and $x'_6 = \sqrt{-1}vt(z')$ are selected instead of the real model coordinates for the time $t(x) = x_4$ and $t(x') = x'_4$ and $t(y) = x_5$ and $t(y') = x'_5$ and $t(z) = x_6$ and $t(z') = x'_6$, then also in the general-special theory of relativity, space-time is generally specifically determined as a universe according to Euclid with six dimensions (Section 45) (Einstein, 2009), as long as the curvature g_{ik} as the seventh dimension is not taken into account with regard to a non-Euclidean universe. A seventh dimension can therefore be added to these six dimensions to take into account the geometry g_{ik} (optics dynamics) of a non-Euclidean universe in the form of a curvature angle coordinate α, whereby the space-time continuum of general special relativity is completely described as a non-Euclidean and yet Euclidean universe.

56 On the general-special relativity theory confirmed by experience

Every experience to date that has confirmed the general and special theory of relativity by Albert Einstein (2009) also confirms the general-special theory of relativity as a consequence of looking back from the general to the special to the general-special theory of relativity. This concerns the perihelion motion of the planet Mercury on its elliptical orbit around the sun, the redirection of light motion due to the gravitational field of a star, as well as the red change of the light spectrum that was confirmed by Hubble for our universe, which is why reference is repeatedly made here to section 3 of this book. In addition, reference is made to all confirmations in the literature that recognize the general and special theory of relativity as a whole or in individual components. In the case of critical opinions towards the general and special theory of relativity, it must be pointed out once again that the general theory of relativity in particular requires a certain effort to understand the model tensor calculation (vector calculation), because only in this way could quantum mechanics be combined with the general theory of relativity to form a quantum gravitational field theory.

All descriptions from the theory that have not yet been accessible to practical experience, regardless of whether they concern the general, special or general-special theory of relativity, which may still be or will probably never be possible to verify (for example, assumptions about the behavior of the edge of our universe), can nevertheless always be evaluated. For as long as these assumptions are conceivable and do not contradict other theorems (axioms), a certain truth must be generally accepted. This is directly justified by theory formation, because this must always be based on conceivable physical facts; something that is not conceivable may not even be included in such a theory. To test theories of relativity, experience should therefore also always be generated in relative terms: For example, via experiments on two different celestial bodies (such as the Earth and the Moon) and other space-time observations carried out several times (simultaneously) on one celestial body, whose results obtained independently of each other can be viewed as documented snapshots using various metrics (space-time rods). In this way, experience with regard to previously unexplained phenomena from the physical reality of humans should be directly transformable into new knowledge about the physical reality of our universe, since in general, according to this scientific-philosophical view, every knowledge of humans also directly represents an experience of humans and since optics dynamics geometrically substitutes (or renews) a purely formal logic and mathematics in accordance with our nature and perception; the optics dynamics of (at least) two snapshots and the light movement taken into account in them therefore represents a completely (new) real logic and mathematics according to physics. For if I stand at rest in the center of a moving circle, an observer sees the rotating planetary surface (of our earth) from above, and imagine two uniform curvilinear hexagons like constellations above this circle in the night sky in order to understand the light and the rotation of the fixed stars as coincident with nature, then I obtain the true Pi of the universe (π_U) as a more accurate Pi (π) with the help of the distance of the twelve fixed stars to the circumference of the circle.

57 The spatial structure in connection with the general-special relativity theory

In order to be able to resolve a serious lack of agreement between theory and observation, i.e. the limitation of Hubble's observation, not only for quantum physics, it must be understandable how this incongruence can be compensated for with systems of equations (Einstein, 2009). The problem here is the lack of experience, because astronomically and physically we can still only observe and understand our solar system, while we can observe all external star systems, but not necessarily equate them with our solar system, so a direct understanding (or an immediate truth) probably seems to be wrong (everything else is an illusion), whereby the deviations in the formation of our universe between the Hubble observation and astronomical physics would have to be explained as a short versus a long period of time (Einstein, 2009).

Is there possibly a paradox between Hubble's observations and the understanding of relativity based on them? Hubble's assumptions are based on various snapshots of galaxies, to which he attributed different lines and changes in these lines. Although this seems possible from an intellectual point of view, it is physically questionable because the gravitational fields prevailing in the photographed galaxies appear to be indeterminable to this day due to the distance to our solar system (apart from mathematically inaccurate approximations), therefore these straight lines should actually be geodesics and their changes should be determined not only in length but also by their angles in arc seconds, which could at some point prove the Hubble assumptions to be too inaccurate, since spectral line shifts should not allow any direct conclusions to be drawn about gravitational field changes in the space-time regions under consideration – at least not in a constant manner. The Hubble observations, here I agree with Albert Einstein (2009), therefore only prove the red shift of light caused by the growth of our universe as a whole (Einstein, 2009). More statements about the relativity of our universe are still not possible today, since experience on site at any place in our continuum is still unthinkable or at least experience in a larger radius around our solar system is not available, which means that a higher accuracy of Hubble's analyses also seems inconceivable due to his snapshots, although observations can never be deceptive or directly wrong, only the understanding gained from them can be inadequate.

(Because) all measurement data of observations (or their empirical research results) accumulated by astronomy, detached without equation systems, offer no basis for deciding whether an expansion of space or its general state appears infinite or finite (to three dimensions), regardless of originally mathematically determined spatial states that allowed a closed space as a result (Einstein, 2009). This is justified by the fact that all empirical research results of astronomy (model observations) must mutually agree with the theoretical research results of physics (model justifications) on the basis of mathematics (model objectification). Consequently, real phenomena and objects in our universe must be observed and empirical data collected about them, which, for example, presuppose the gravitational field equations for the model justification of the relativity of our universe, thus making the connection between astronomy and physics apparent. The concept of astronomical physics originates from Albert Einstein (2009), but this consideration has not been explored in such depth as follows: Therefore, no physics without astronomy (physical astronomy) applies, because the observations must always (as far as possible) resemble the justifications with regard to the model of our universe, just as no astronomy without physics (astronomical physics) applies, because the justifications must always (as far as possible) resemble the observations with regard to the model of our universe. Only now can "more precise statements" about the relativity of our universe or about a shorter or longer period of time for the formation of our universe as a whole gradually become conceivable, since the theoretical equation systems of physics must then agree "with a smaller deviation" with the empirical data systems of astronomy in both directions by substitution between a starting point and a (later) event state (movement) or end point and an (earlier) event state (development). For theory and practice in this significant fusion are psychologically and mathematically more "nature-coincident" and must therefore, according to truth, lead to more "covariant" general laws of nature, which are also more "law-coincident" according to the theory of relativity and thus describe the existence of our universe in a more "physically real" way.

58 Space theory and its relativity in general-special

General-special gravitational field theory. An assumption about an independent and yet dependent gravitational field is not difficult to make on the basis of general-special relativity theory, because a "fieldless" Minkowski space state must coincide with a certain metric property of the general gravitational field theory (Einstein, 2009). With the help of this special case, the extended gravitational field theory arises as a consequence of a generalization that is almost exclusively systematic (Einstein, 2009). The physical reality of our universe (continuum) is to be understood in the same way as a light field; this light field resembles a generalization of the gravitational field. The light field theory resembles a generalization of the theory of an independent gravitational field that is nevertheless dependent on a light field (Einstein, 2009). This generalized-special gravitational field theory would have to remain confronted with all empirical possibilities (of astrophysics) (Einstein, 2009), except for assumptions about his quantum mechanics, because this does not represent a theory about the quantum gravitational field, since such a theory must combine quantum physics with astrophysics to form quantum astrophysics.

A generalization can generally be described as follows (with regard to a light beam that describes the field of gravitation as a field, i.e. even more generally-special) (Einstein, 2009). An independent light field of all $L_{ik} = G_{ik}$ has a symmetric function L_{ik} equal to L_{ki} (L_{34} equal to L_{43} etc.) in accordance with the contentless "Minkowski space state" (Einstein, 2009). A light field exists in the general case with the same properties, but with an anti-symmetric function L_{ik} not equal to L_{ki} (L_{34} not equal to L_{43} etc.) (Einstein, 2009). A deduction (derivation) of the light field theory is completely similar to the special case of an independent light field (Einstein, 2009). The following system of equations (83) therefore corresponds to the general-special gravitational field theory via the covariant general-special laws of nature.

(83) $ds^2 = g_{ik}dx_{ik} = g_{ki}dx_{ki} = L_{ik}dx_{ik} = L_{ki}dx_{ki}$

With regard to the previous general consideration, a question concerning the quantum mechanical field theory of light and gravitation appears to be of secondary importance (Einstein, 2009). One of the most important questions today is whether the light field laws can simply lead to a truth with regard to their considered nature (Einstein, 2009). Here I am referring to light field theory, which can fully represent physical reality (taking into account a spatial state with at least four dimensions) with the help of a gravitational field (Einstein, 2009). The current crop of quantum physicists tends not to answer this (most important) question with a yes; this crop of quantum physicists believes, as a consequence of the current expression of the general gravitational field theory of Albert Einstein (2009), that a state of a space(-time system) is by no means directly describable, but instead only directly describable by means of mapping numerical statistics of all measurement results that can be experienced within this space(-time system); the prevailing view is (mentally) limiting, as if a tentatively evaluated trialistic light nature (body wave field design) could only be experienced with such a described minimum level of understanding of reality (Einstein, 2009). According to Albert Einstein (2009), it can generally be assumed that such an extensive group-dynamic rejection of knowledge continues to exist without sense and reason despite today's (single-dynamically equal) knowledge experience, and that the general public (of quantum physicists or all disciplines together) must think themselves challenged by this to further develop this method for deriving a theory of relativity via the light field to completion (completion), i.e. to modify it with regard to the quantum gravitational field (Einstein, 2009). For such a theory can always be established directly in connection with the state of any space-time system (subjective deduction), i.e. without the need for numerical statistics of measuring physics (objective induction), since experience can be learned in at least two fundamental ways: perception (observation), then understanding (objective induction equals experiential knowledge) or understanding, then perception (observation) (subjective deduction equals cognitive experience). Here, based on the present book on general-special relativity theory and its derivation via general and special relativity theory, subjective deduction proves to be the more efficient way of thinking for generating knowledge analogous to experience specifically with this unified science, which I have termed business informatics physics, thus also generally with a science of everything unified from arbitrarily selected subfields.

Reference

Einstein, Albert (2009). *Über die spezielle und die allgemeine Relativitätstheorie*. Springer, 24. Auflage.

FSC
www.fsc.org
MIX
Papier aus verantwortungsvollen Quellen
Paper from responsible sources
FSC® C105338